GEOPHYSICS

GEOPHYSICS

Annette Bolger

Oxford Book Company
Jaipur ▪ New Delhi

Published By

Oxford Book Company

Regd. Off. "Bony Residency", Gate No. 2
Opp. Tilak Public School, Vishweswariya Nagar,
Gopalpura Road, Jaipur - 302018, India
Ph: 0141 - 2761280, Telefax: 2761381
email - oxfordbookjpr@gmail.com
website - oxfordbookcompany.com

Delhi Office: 102, 1st Floor, Satyam House,
4327/3, Ansari Road, Darya Ganj,
New Delhi - 110002, India
Ph: 011-45652440

Published in India, 2024

ISBN : 9789380179452

Price : ₹ 2225/-

Typeset by Shivangi Computers, Jaipur

Printed and bound in India by Thomson Press Ltd.,
New Delhi

Preface

Geophysics, the study of the earth through the application of physical principles, first made its appearance in the mid nineteenth century and progressively influenced our knowledge and understanding of the planet we live on. In turn, this opened new horizons of comprehension of the complex natures of the continents and oceans as well as the natural resources available to us, on which the growth and development of our society is based. This all-embracing science is a fundamental pillar on which global socio-economic prosperity is based and it is indeed difficult to imagine today's world without its benefits.

Geophysics helps us to unravel the mysteries of the unreachable insides of the earth. It has made unparalleled contribution towards knowledge of lithosphere structure of the Earth and the Plate Tectonic processes, which over millions of years, build the surface features and amass mineral wealth for our consumption. Understanding of these as well as other processes acting deep inside the earth enable us to anticipate regions prone to earthquake hazards as well as other vagaries of nature. Again, knowledge of the earth system and man-made influences on it have helped to open our eyes to activities, which generate pollution and create imbalances endangering the very existence of our species, so that we may implement rectifying measures.

Annette Bolger

Contents

Chapter 1

Geophysics

Geophysics is the study of the earth's properties by applying physical theories and using instruments for measurement. Since antiquity men have been studying the earth's properties in order to predict earthquakes, but real progress happened in the 1500's and has continued since then. Scientists began working to understand how magnetism and gravity are related to earthquakes. Magnetism is an attraction to metals such as iron, nickel, cobalt, and their alloys. Its main character is a magnetic force. Though metals are more strongly influenced by magnetism all materials are under its influence to a degree.

One of its particular manifestations in physics is that it's a component of electromagnetic waves. Light is an electromagnetic wave. Magnetism is now used to help predict the frequency of earthquakes by utilizing the fact that earthquakes happen in clusters. They can't yet use this information to predict exactly when an earthquake will occur, but when there is an earthquake statistics show others will follow. Heat is a factor in magnetism and when the heat in an atom is at an intermediate temperature according to "Curie temp," then instead of the clusters of atoms moving in different directions like when under high temperature, aligned clusters form and these groups move in the same direction, though there is more than one grouping.

Thus, heat is one of the properties that cause earthquakes and this is also related to gravity. That is when earthquakes occur they change the local gravity. Massive sections of the earth called continental plates move and meet at fault lines and cause the change in local gravity. Besides the study of magnetism and gravity these are some of the other areas studied in geophysics: lithospheric plates, refraction seismology, crustal roots, ocean basins, isostatic equilibrium, and subduction. In more depth the study of geophysics is an interdisciplinary effort involving Mathematics, Physics, and Geophysics in the course load for students. Geophysics is divided into two large parts: Exploration Geophysics and Global Geophysics. They study the structure and composition of the earth. There are many subdivisions of geophysics including: oceanography, atmospheric physics, climatology, petroleum geophysics, environmental geophysics, engineering geophysics, and mining geophysics.

What other work do geophysicists do? Usually they are part of a team with other earth scientists like geologists and engineers. They work on environmental, construction and exploration problems. They use instruments and computing to for measuring physical properties. They measure density, electrical resistivity, electrical fields, radioactivity of rocks, velocity of sound waves transmitted through the ground, changes in gravity and magnetic fields of the earth, and reflection of radio signals coming from rocks near the surface of the earth. What do geophysicists use these measurements for? They use these measurements to find oil, natural gas, metals, diamonds, potash, and coal.

They also identify environmental problems and to calculate the qualities of an area, judging how suitable they are for dam building and construction sites. Geophysicists get jobs working in consulting, oil, academic and government employment. Geophysics is an exciting field of endeavor that properly utilized can improve our living environment and teach us about our home—the earth.

AIRBORNE GEOPHYSICS

Airborne geophysical systems have been in use since 1943. They used magnetometers for geophysical mapping and detecting submarines. The airborne electromagnetic systems came out in the 1950's. These early devices were mounted on airplanes. In the 1960's these systems were applied to use with helicopters. Airborne geophysics utilizes several types of surveys to accomplish their goals like airborne magnetic surveys, electro-magnetic surveys, and radiometric surveys. The three main airborne geophysics procedures are magnetic, electromagnetic, and radioactive.

These airborne methods have a big impact on the mining industry. Better geophysical tools, global positioning systems, and airborne platforms have been useful in engineering and environmental problems, though only to certain problems in these two areas of endeavor. Airborne geophysical surveys are also used for oil and mineral exploration and land management.

Airborne geophysical methods have been used in detecting and mapping exposed bedrock. But better methods are needed to explore natural aggregate resources. Magnetometer surveys map magnetic occurrences in the earth's magnetic field caused by magnetic minerals earth's upper crust. Electro-magnetic surveys are aimed at conducting a quick and cheap search for metallic conductors. It is also applicable to general geologic mapping and many engineering problems. The radiometric surveys look for radioactive material in rock. Airborne geophysical surveys are used for land management as in Western Australia where huge parts of the farming land has been affected by soil salinity. Geophysics has shown that topography and other factors are the cause of the high salinity of the soil, according to The Bulletin of the Australian Society of Exploration Geophysicists.

Airborne electromagnetic surveys are useful in investigations of different geological eras. For instance, they can help in locating subglacial valleys, which helps to understand water reservoirs. There are still many unanswered questions about the geologic structure of Antarctica and these questions may be answered by airborne geophysics. Maud Land has proved that airborne geophysics is capable of helping to map Precambrian Antarctica. Antarctica plays a key role in climatic and geodynamic history of the earth and again airborne geophysics can help learn about this history and the climatic role involved and how this data affect the whole earth. Climate research is enhanced with airborne research, specifically studying ice sheet and lithosphere interactions. They do this by using multiple sensors.

Scientist are trying to determine how these two physical factors interact with each other causing long-term climate variability. In these studies they use ice-penetrating radar on long-range research aircraft. They have to have special antenna structures that can operate at 300 knots. They first tackled this problem in the 1970's working in Antarctica. Airborne geophysics has the capability of being used for a myriad of uses including mining, solving environmental problems, engineering, land and water management, archeology, detecting submarines, and many other scientific applications. With global warming and other atmospheric problems it seems that airborne geophysics can play a vital role in solving them and making the world safer for everyone to live in and it can help solve our energy problems.

GEOPHYSICS INSTRUMENTATION

What instruments are used in geophysics? The instruments used in geophysics are seismographs, land and water magnetometers, and geoelectrical instruments. The different types of seismographs are the channel seismic recorders, seismographs for refraction, depth to bedrock teaching and groundwater studies, high-bandwidth digital marine seismic streamers, seismic recorders for analog streamers and software that processes reflection, refraction, surface wave and marine data. What are geophysics instruments used for? They are used geological studies. The studies concern rock and soil studies and surveying the earth. The subdivisions of geological study are geophysics, geochemistry, mineralogy, hydrology, and geomorphology.

Geophysical instruments like seismographs and magnetometers help in the study of earthquakes and surveying the earth's crust. The function of a seismograph is to detect and record the intensity, direction, and duration of an earthquake. From this information scientist can calculate how much energy was released and to pinpoint the source of the earthquake. A magnetometer measures the strength and direction of magnetic fields. The two fundamental kinds of magnetometers are the scaler magnetometer and the vector magnetometer. The first type measures the total strength of the magnetic field. The second type measures the components of a magnetic field in a certain direction. Magnetometers can measure the magnetic pull of metals.

Therefore, they can locate iron in ships and other objects. They can be used to find wrecked ships, archeological sites and buried treasures. About 12 years after WWII gains in the geophysical apparatus used for the purpose of prospecting oil. The specific success was in developing an instrument that could accurately register anomalies in the earth's crust from an airplane. Improvements in geophysics instrumentation keep moving ahead and from 1970-1974 advances were made in marine geophysical instrumentation systems. And innovations in different areas of geophysics instrumentation continued so that these days digital models and computer data are used. Geophysics instruments are used in the study of archeological geophysics.

They have taken the geophysical used in mineral exploration, engineering and geology. The main instruments used are magnetometers, electrical resistance meters, ground-penetrating radar and electromagnetic conductivity meters. Electrical resistance meters are useful when archeological features have a higher or lower resistivity than surrounding ground or rock formations. Electromagnetic conductivity meters work in a similar manner as electrical resistance meters, but have some advantages. They don't need direct contact with the ground and they will work in conditions that electrical resistance meters won't work in. Most of the time in archaeological endeavors the gradiometer configuration of the magnetometer is preferred to the use of one sensor because it gives better picture of small objects on the surface of the ground.

For the use of the ground penetrating radar sand soils are perfect. In this use a radar signal is aimed at the ground subsurface objects and layering reflections are detected by the receiver. The amount of time it takes the signal to travel is an indication of depth of the object detected. Although this article doesn't cover all of the ways in which geophysics instrumentation is used by geophysicists, it should supply an idea of the wide range of uses of the instruments.

GEOPHYSICS VERSUS PHYSICS

When people talk about geophysics they are sometimes confused with the discipline of physics. Geophysics is a branch of Earth science. These scientists study the Earth by quantitative physical methods and use seismic, electrical, magnetic, and radioactive methods to prove theories and to actually do work in the field finding resources such as gold, diamonds, and other treasures of the earth yield beneath the soil. People who work in the geophysics field work with oil companies and other mining endeavors that produce resources for the global economy. Geophysics is paid on a salary basis for sometimes or giving a percentage of the yield.

It is an exciting end demanding discipline that requires a massive amount of math and science classes in high school and college. Jobs in the geophysical realm include atmospheric science. These scientists study the terrestrial magnetism and atmospheric electricity. Their studies may also cover climatology, meteorology, and radiant energy. Other fields within geophysics include galcierology which the strata's of ice are studied from the melting glaciers, hydrology in which the study of the surface and groundwater is made important, and geodynamics which is a numerical study of the error. All of these fields require a firsthand knowledge of sophisticated computer software and data processing units. There is extensive fieldwork in which the geophysicist will go outside in look at the stratus of Earth and the composition of the soil to determine what kind of resources lie underneath.

The other half of the day is spent taking those observation components into statistical models on the computer to prove their observations mathematically. The difference between geophysics and physics is that physics is a science of matter and motion. By taking space-time and other considerations of the physical world works, physics uses experimental sciences and general analysis of nature and how the Earth interacts. Geophysics is one of the newer sciences that have derived from physics while physics itself as one of the oldest academic disciplines. The combination of mathematics and philosophy are also an important part of physics that contribute a theory or law into the realm of the earth and its motions and workings.

When we look at physics versus geophysics we will find that the physics realm is the base science in which geophysics derived from. Though they are interchangeable in certain attributes, geophysics has evolved past the normal physics discipline and can be considered a stand alone science. Geophysics can use the tools and the discoveries of physicists in their endeavors to find natural resources within the Earth's crust. This is because a physicist work is derived from mathematical reasoning and put into technological advantages.

Work in physics as well as geophysics is both experimental as well as theological nature. The real world applications of physics and geophysics to the technology sector and other industries provide a basis for every material thing that we as humans use every day. The study of mass and motion compiled with the direct study of geophysical activity within the earth's crust of atmosphere as a combination that can develop new approaches to environmental, ecological, and physical problems that exist in the world today.

GRADIENT GEOPHYSICS

After the Grand Daoui horizon was exhausted, the great Oulad Abdoun Basin exploitation of a new phosphate deposit began. Some so-called 'disturbances' such as sterile hardpan are hard to detect and can interfere with the phosphate extraction. The geophysical surveys indicated several prospective anomalies within the possible Paleozoic basement rock. Representing more than 75 per cent of the world's reserves in phosphate, Morocco has an annual output of 19 million tons and a reserve in excess of 35 billion cubic meters. In the Sidi Chennane deposit, it is sedimentary stratiform and contains several distinct phosphate-bearing layers.

Found in contact with alternating layers of calcareous and argillaceous hardpan, these layers, in this field were extracted after Grand Daoui deposits were exhausted. Locally known as "derangements' or disturbances, the new deposits contains many inclusions and lenses by extremely tough hardpan found throughout the phosphate-bearing sequence. Normally, hardpan pockets can only be detected at the time of drilling. They interfere with field operations and introduce a severe bias in the phosphate reserves. Well logging or surface geology – direct exploration methods – is not particularly effective. Although, the chemical changes which are detectable at the hardpan and phosphate rock interface produce an important resistivity contrast.

Factors such as changes in clay content and lithofacies and the consistence appear to account for some additional resistivity difference. Located within the Oulad Abdoun basin, the Sidi Chennane phosphate deposit is about 33 kilometers South-East of Khouribga. West meridian 372500 – Lambert – South meridian 22800 – Lambert – East highway RP22 and North are its boundaries, having the outcrops of the basement of the phosphate-rock sequence. Vegetation is mostly sparse dwarf palm trees due to the areas arid climate of phosphate plateau. November to May is the rainy season and it's usually kept below 400 mm. The rural population exists on ranching cattle and seasonal agriculture in the small villages or dour.

Ground water is increasingly in short supply. The geologic section rest uneasily on Paleozoic schist and quartzites and the basement is well sedimentary covered in fairly thick Maestrichtian and Eocene which contain the phosphate-bearing strata which are 30 to 50 millimeters thick. Below the phosphate-bearing strata, we can find up to 70 meters of Senonian marls and limestone marls; 20 to 60 meters of Turonian limestones; a Cenomanian formation of alternating limestone marls and gypsum marls and finally, 10 to 60 meters of red marls and mudstone of pre-Cenomaniage age. Differentiated by size of the pocket or inclusion, the disturbances may be a type of material, clay content, hardness or type of contact with the phosphate rock.

There are usually two main types of disturbances found. The first type is throughout the mineral deposit and it appears to be a random mixture of limestones, clays, cherts, marls and a low-grade phosphate with large amounts of cherty limestone. The second type is a highly disturbed and lacks any dominant facies. It appears as an accumulation of low-grade phosphate limestone blocks with large nodules of marl, chert, some fragments of chert and phosphate rock.

OIL AND GAS EXPLORATION

Oil and gas exploration come under the group of petroleum geology, this is their scientific designation. The search for hydrocarbon deposits under the earth's surface is hydrocarbon exploration or the hunt for gas and oil. Different forms of petroleum have been found and used by various civilizations for thousands of years. They found oil seeps and other visible surface features like natural gas seeps and pockmarks. Pockmarks are underwater craters caused by escaping gas. These are evidences of hydrocarbon generation, which may be shallow or deep in the earth. Today the need is much greater and sophisticated methods are employed to find and estimate the amount of the deposits. There are several geologic elements that are required to make deposits large enough for exploration. There must be an organic-rich source rock which generates the oil or gas.

Rock porous enough to be the reservoir rock that stores the petroleum. A trap that stops the petroleum from leaking off There are certain places where traps are normally formed. The tops of anticlines. By faults the updip of pinchouts of sandstone beds beneath unconformities. When petrophysicists think an area contains oil or natural gas they subject it to a text like a gravity survey or a magnetic survey. This helps them to detect larger features of the sub-surface geology. These are features that are interesting in terms of exploration. Today they use large vibrator truck to create the shock waves; historically they exploded small dynamite packages in shallow holes. There are environmental laws these days that prevent using explosions to collect seismic data. Seismic surveys are taken of leads.

They send shock waves into the ground. Seismic surveys measure by calculating the time it takes reflected sound waves to go through rock and be reflected back to the surface. The time it takes is related to how dense the rock is. With this information they use the process of depth conversion to create a profile of the substructure. They detect the returning sound waves with geophones. They then process the geophone data with computers converting it into seismic lines. These are two-dimensional displays that look like cross-sections. They used to be two-dimensional lines. They create them by laying the geophones out in a single line. Now they use 3-D seismic volume. Which they use to make 3-D computer models of the subsurface geometries of the rocks. So petrophysicists are hunting for rich oil fields.

They are looking for a reservoir shaped to trap hydrocarbons that is covered by a sealing rock. Besides oil and natural gas there are forms of petroleum called asphalt and tar. These are solid and semi-solid forms respectively. The name for dark, viscous liquid petroleum is crude oil and the clear and volatile called condensate. But even with all of this sophisticated equipment they don't always find an economically viable oil deposit. The development or extension has the best chance for success. Wildcat wells have a lot less success at finding a good oil well.

PETROPHYSICS IN BRIEF

Petrophysic studies the physical and chemical properties of rocks, soils and fluids. These properties describe the occurrence and behaviour of these elements. It is a subfield of geophysics. The properties that geophysicists study are porosity, density, magnetization, electrical conductivity, solid mechanics, thermal conductivity and radioactivity. That said, they study ore deposits and oil and gas reservoirs. They work in the oil, gas, water and mining industries. They measure and compute conventional petrophysical properties, rock mechanical properties and ore quality. The main concern of petrophysics in the oil and gas industry is the behaviour of fluids in rocks.

They are also concerned with pressure. They measure these properties (resistivity, natural radiation, sonics) during and after the drilling of an oil well with special instruments. The make these measurements down in the hole. Sonics is the speed of sound. With these measurements they can estimate the type of rock and if oil and gas are present. To make accurate characterizations of an oil or gas reservoir they take measurements of resistivity, density and measure neutrons. From these measurements porosity, saturation and permeability are calculated. They quantify these properties. Why do they need to quantify these properties?

This is how they study an oil field or gas reservoir. That is how they know how much oil is there and basically where it is. Sometimes a fault or dome stops the oil or gas from spreading. These understand how geologists study these elements an understanding of the properties is in order. Porosity is the volume by unit of an oil formation. The porosity of a formation can vary quite a bit. Dense carbonates can have a very low percentage of porosity. Limestone and evaporates are dense carbonates. Shales and clay can have up to 40% porosity. The problem is that the individual holes are usually so tiny that the rock is impermeable for fluid to flow through it. Consolidated sandstones vary from 10% to 15% porosity. Non-consolidated sandstones porosity can be as high as 30%.

Water saturation is a factor in determining whether there are hydrocarbons in a rock formation or not. It also helps to estimate the amount of existing hydrocarbons in a known reservoir. This is the calculation of the reserves. Permeability is basically how much liquid a rock can absorb. This is how the rock's ability to transmit fluids is measured. For a rock to be permeable it must have interconnected fractures. These are also called pores. Thus a relationship between permeability and porosity exists. Another tool used in oil exploration is the petrophysical model. It's used to determine reservoir characterization. That is procedure with which to interpret petrophysical data. This data normally involves equations, algorithms, or another mathematical process.

Many times these models have multiple routines. These are usually wireline logs. There are different types of models. An example is a deterministic model and its calculations include: the shale volume, total porosity, effective porosity, water saturation and permeability. Usually the model is calibrated using core and production along with other data sets. Although these models are part of many software packages, unique log-analysis problems are common and a model they need their own special models built in order to make correct calculations.

PLANETARY GEOPHYSICS

Planetary geophysics is the study of the earth's core, the mantle and crust, to the outposts of our solar system and beyond. We begin with our own planet and studying earthquakes and their formations or the lack of movement in some areas where there is so much movement in other areas of the world. By studying this, we can determine why one area of the planet tends to have more earth quakes more than another area and what might prevent them in the future, saving lives and countless amounts of money in damages. The planetary geophysics studies the earth down to its core and makes discoveries about the natural gases and lava and other items buried beneath the surface of the earth that can one day be helpful to us.

The planetary geophysics includes the study of other planets such as Mars and Saturn and their various conditions verse our own atmosphere and conditions. There is also the terrestrial geophysics or studies – the search for other beings on other planets. And the study of remote environmental changes. There are many schools and centers where one can go and study planetary geophysics and graduate with a degree in planetary geophysics. Jobs in the planetary geophysics include those with the space programme, studying the planets for activity and changes in their makeup. Or helping to launch the space ships and helping to gather and catalog what they bring back with them. Another job in the planetary geophysics is the study of our planet, earth.

This could take on a wide range of jobs including a seismologist, one who studies earthquakes; or someone who helps with environmental and global warming issues such as the arctic melting. Without our planet, we would have no other planets to look at and study so beginning with our planet might be a good place to start. However, many people believe that it might be too late for planet earth and that we should be studying other planets incase we can no longer live on this planet. Where would we go? This is a question a planetary geophysicists could answer for us.

There is so much research to be done by planetary geophysicists and scientists that can work with the planets and don't forget the moon and sun for they are planets as well. Scientists are studying the sun and moons impact on our planet and that of other planets. We know that the sun plays a major role in our everyday lives, but what about the moon? What role does it play in the average person's life on planet earth? Planetary geophysicists study and discuss these and many other questions so that we may one day live in space or on the moon if we have done too much damage to planet earth.

Which, some say we already have with global warming and the pollution of many factories and just everyday carbon emissions from people doing everyday things like driving. Or the massive landfills that will soon need to be made bigger because no one is recycling, which is so important today.

STUDY OF ARIZONA

The depositional process uses well-rounded gravel, course gravel of the Mogollon Rim in Northern and Central Arizona, called rim gravel which was examined at two separate and widely respected locations for this study. Literature gave further information regarding the rim or course gravel. The gravel is very course near the East-Central portions of Arizona and the course gravel is occupies the uppermost terrain in the region. It has been found that the course gravel was deposited as a sheet it eroded into small pieces during the Recessional Stage of the Genesis Flood.

A consensus was met that the Rim Gravel provides the facts that the Flood and post Flood borders match up with the stratigraphic location of the rock which has been named "late Cenozoic" in this uniformitarian geological column in the Western part of the Untied States. Many formations of geomorphic features, including such wonders as the Grand Canyon of the Colorado River which makes this interpretation relevant to these theories. Gravel that is coarse on top of the Mogollon Rim in the Northern and Central Arizona is called the rim Gravel and has a great significance for questions of historical geology in the American Southwest.

The coarse gravel occupies an erosion surface on the highest terrain in East-Central Rim. The course gravel occupies an erosion surface on the highest terrain in the region and is believed to have once been continuous along the Rim. A large percentage of the gravel is an exotic quartzites, sometimes with percussion marks. Based on literature sources, paleocurrent data indicate the coarse gravel was transported from the west and the south that currently is at a much lower elevation than the Mogollon Rim. Some of the Paleohydrological analysis has given us calculations that the coarse gravel was transported by sheet flow moving at a rate of at least a few tens of meters per second – about 40 mph or greater. However, the uniformitarian age of the gravel is generally believed to have been early 'Cenozoic," it can be surmised that the gravel and the "Mogollon Highlands" to the south were eroded probably in the mid "Cenozoic" era.

If the erosion for the area is more channelized, it would probably be assigned to the late "Cenozoic" era. This is based on the assumption that these uniformitarian classifications have any real meaning at all. We conclude that this coarse gravel was deposited as a sheet during the early Recessional Stage of the Genesis Flood. The area then underwent erosion of the deposited gravel and substrate during uplift of the area, generally during the Channelized Phase of the Deluge. We can then infer that the Rim Gravel provides evidence that the Flood and post-Flood boundary largely corresponds to the stratigraphic location of rocks termed "late Cenozoic" in the uniformitarian geological column in this part of Western United States. And since the Grand Canyon of the Colorado Rivercuts through the Rim Gravel, this characteristic must post-date the deposition of the Rim Gravel at least slightly.

A DEPOSIT CORE ANALYSIS A COMPARISON IN METHODS

A deposit core analysis is a method in which the geophysical discipline takes samples from strata of earth, rock, or soil. There are several types of core analysis which vary in degree in accuracy. One of the most intricate systems for core analysis is the core gamma core analysis that delivers the results in a total and spectral realm through the use of gamma radiation. There are even portable gamma core analytical devices that allow the scientists to test the density of the core by counting the y flux attenuated by the core. This allows for more diversity in the field and allows for more accuracy in the depot core analysis. Another type of depot core analysis is porosity and permeability.

The porosity is a measurement of how much open space there is in a rock. The space measured during core analysis can be between grains in the rock or between cavities or breaks within the rock or core. The permeability is another analytical measurement that corresponds with porosity by testing how easy water can move between the before mentioned cracks. The variance in pressure in which the water is forced through the cracks or grains can be as low as 1 psi and as tremendous as 10,000 psi. Slabbing is another technique that delivers a radial or band saw result that will allow the core analysis to take on an interior look into the rock or soil.

This is the most popular core sample analysis technique because it allows the researcher or scientist to analyse the intricate by looking them directly or to study using enhanced digital imaging. The only drawback to this method is that some cutting core analysis machinery leave oil residue that could deflect from the true readings and validity of the sample. New techniques in lubricant and coolants have reduced the chance of sample damage. Sometimes there are more extensive and intrusive sample analyzing that cause for a full diameter core analysis. Here rocks such as carbonates or vulgar rock can be difficult when plug type small analysis is used. A deeper and more all inclusive core analysis is needed for this type of rock.

The difference between core plug, full diameter, and whole core analysis is the length of where the sample is taken and how detailed the analysis will be. The core plug will separate a two foot core sample into six inch sections and then a one and one half inch plug is removed. The full diameter core analysis will take the six inch sections of the two foot piece, but does not extend to taking samples from those pieces. The whole core analysis is the entire 2 ft core sample as the sample itself. A core sample can be use d to determine the conditions of the earth and the rock when the rock was formed or was on the surface of the earth. Each core sample analysis can yield the type of flora and fauna the earth held and what kind of atmosphere existed during that time.

SEISMOGRAM MEASURING THE EARTHS MOVEMENT

When the earth trembles, as it does during an earthquake, a seismograph takes a reading and the resulting measurements and readings is called the seismogram. Many years ago, a seismogram used to be a pencil attached to a revolving drum where the lines would be drawn to represent the activity going on during the earthquake or sometimes an explosion. Looking at a seismogram, we will first notice all the squiggly lines running across the paper. These represent the seismic waves that the seismograph has picked up during an earthquake or some type of movement below the earths surface.

Microseisms are so small that no one ever feels them, they don't cause any damage and they will not hurt anyone. These small 'tremors' can be anything from waves hitting the beach or heavy traffic near the seismograph. These microseisms can also be caused by the wind or any number of ordinary activities in the atmosphere or man made such as construction drilling near the site of the seismograph. Among the wiggly lines, the first set, after the long, plain and straight line which represents no activity, would be the P lines and this represents the P wave and is much bigger than all the smaller, microseisms wavy lines. Next on the seismograph will be the S waves which are bigger than the P waves.

This indicates the next wave of damaging earth movements. The absences of S waves on our seismogram means that the earthquake probably happened on the other side of the planet and it was not able to be felt or picked up by our seismograph. S waves can't travel through the liquid layers of the earth, therefore, these S waves never made it to our seismogram. Often the larger, longer waves marked on the seismogram are the surface waves, sometimes called the Rayleigh or Love waves. These waves have a lower frequency which means the up and down lines on the seismogram are spread out more. Traveling a little slower than S waves, these surface waves arrive directly after the S waves. If the earthquake was a shallow one, meaning it was close to the surface of the earth, the surface waves will be the largest waves to be seen on the seismogram.

Seismographs and their information – seismogram – can help save lives when they are read by a team of scientist know as seismologists who can predict when the earthquakes are going to be coming based on the latest data.

Once they see the first wave of lines, the P lines, then they can basically predict within minutes when the S lines and surface lines, if there are going to be any, will be coming and warn the public through the proper channels, such as radio or television. Today, a lot of the seismograms have come a long way since the drum and paper and pencil kind of the first earthquake predictors. Today, there are many seismograms that rely on the technology of computers and various stakes and antennas placed in the ground and around the seismograms to help gather the information to help save the public.

SOURCE ROCK EVALUATION

There are many reasons and ways for source rock evaluation methods, to obtain petroleum and other gases or oils and countless ways to extract these items. As rocks are obviously buried, the contained organic matter is heated or matured within the Earth's geothermal gradient, to generate petroleum.

The thermal level of maturity for source rocks is consequently a basic and critical parameter in assessing petroleum prospectively of exploration acreage. They hydrocarbon potential of a source rock comprising insoluble organic materials, capable of generating hydrocarbons upon pyrolysis and regular hydrocarbons or further maturation is analytically determined. In the instant invention, a particulate sample of the source rock is slurred with a solvent to extract the hydrocarbons there form and provide a first solution for analysis.

Then the extracted sample is again slurred with a solvent and the resulting slurry is heated to a temperature sufficient to pyrolyze the insoluble organic material to liquid hydrocarbons. They liquid hydrocarbons dissolve in the solvent to provide a second solution for analysis.

The two solutions are then analyzed independently for hydrocarbon content. Originating from organic material in rocks, petroleum and the capacity of organic matter to have developed oil or gas in the past can be inferred in part from the organic material and its present degree of thermal maturity. Groundbreaking analyses of the Early Cretaceous foundation rocks in the Otway Basin has exposed a thick, mature; organic-rich source rocks containing gas- and oil-prone marcerals exist to the Northwest of Portland in the Tyrendarra Embayment and around the Prot Campbell region.

The sampling opportunities for source rock evaluation are geographically and stratigraphically imperfect because only a small portion of the onshore Victorian Otway Basin is deeply drilled. Biomarker analyses designate that gas-prone marcerals are the most abundant and wide-spread. Oxic, open plain and shallow lacustrine environment conditions prevailed over most of the Northern areas. By dissimilarity, deeper water plummeting conditions prevailed in the Southern and North-Western marginal areas. Progenitor hydrocarbon materials are predominantly asphaltic/aromatic but in some locations, the organic contents contain more paraffinic and naphthenic compounds. In addition, hydrocarbon generative potential ranges from excellent to poor; and the liquid hydrocarbons are present in some of the deep shale's.

Several of the hydrocarbons that were migrated and generated in the Portland region and in the Port Campbell region, were most likely pooled in existing stratigraphic traps. Most structural traps were formed during the Palaeocene uplift event in the Portland region and during the Oligocene uplift event in the Port Campbell region. These structural traps, particularly those in the Port Campbell Embayment, have been loaded by the continuing oil and gas generation and expulsion form the Aptian-Albian source rock.

Despite the lack of state-wide sampling opportunities, analytical results suggest that the most prospective areas for oil and gas – so far as close proximity of reservoir and seal rocks to thick, rich mature source rock is concerned – are within the Casterton and Lindon to Pine Lodge regions, in the Portland/Tyrendarra Embayment and between

Peterborough and Port Campbell offshore and along from the present Victorian coastline. Many companies, for a price, will provide evaluations of petroleum source rocks. They use organic carbon content, hydrocarbon content, kerogen maceral analysis, visual kerogen and vitrinite reflectance.

STRATIGRAPHIC COLUMN EXPLAINED

A stratigraphic column is a drawing by a geologist. The drawing depicts underground rocks or cliff wall rocks, showing their vertical sequence. This is used to interpret the relative sequence of geological past events. It's more like a diagram than a free hand drawing. When another geologist studies the drawing he knows what the rock formation looks like and he can understand the geology. Thus, he has the knowledge to speculate on the possibility of oil or natural gas deposits existing in said rock formation. Simplified stratigraphic columns are used to find oil deposits to drill.

Drawings of stratigrapic columns can also be used for engineering purposes. Engineers try to put the largest amount of information in the smallest space when making geologic maps. By drawing in engineering and hazardous identification concerns they can adaptthese columns to fit their needs. The drawing of a specific area shows its major units of rock. The type of rocks the units are made of is also illustrated. It depicts this information by naming each unit and by what symbols are used to tell the rock type. The density is also shown. This branch of geology is called stratigraphy. It's the study of rock layers and layering. It involves two subfields: lithostratigraphy and biostratigraphy.

Nicholas Steno started the theoretical basis for this subject in 1669. He wrote about fossils found in rock stratification. Later, William Smith created the first geologic map of England. He worked on this during the 1790's and the beginning of the 1800's. Lithostratigraphy studies the changes in the layering of rock types in a formation. Biostratigraphy studies the fossils in strata and relates the fossil content in different stratas from various rock formations that are far apart to each other endeavoring to correlate them in time the fossil fauna and flora compared must be of the same type. There is another branch of stratigraphy called chronostratigraphy. This branch tries to comprehend the geologic history of rocks and regions.

At first, geologic time was measured by the breaks in stratigraphic column. Scientists thought that world wide mountain building caused the breaks. They thought that during this mountain building sedimentation did not occur, therefore the gaps formed. They have realized that mountain building events aren't necessarily universal. They have concluded that mountain building can occur in one continent during one interval of geologic time. Geographic time shows the relationship between time and rocks. The normative stratigraphic column is the base of the geologic time scale. There are four eras in a geologic time scale.

Three of these eras are named for fossils found in the related strata. These eras are the Paleozoic, the Mesozoic, and the Cenozoic. They refer to ancient life, medieval life and modern life. The oldest and fourth era normally lack diagnostic fossils, It's called the Precambrian era. Geologists divide these eras into periods of time. Rock deposits formed during a particular time period make up a system of rock. The periods are branched into epochs. Each rock system is branched into a series.

STRATIGRAPHIC GAPS AND STRATIGRAPHIC TRAPS

If we heard that oil is located in underground pools or lakes then we are misinformed. Most oil is found in minute spaces in sedimentary rocks like sandstone and course-grained limestone. Think of the sandstone as a sponge. The sponge has holes that can contain liquid or gas. Though limestone and sandstone are hard rocks, they are very porous. The holes are very small. Oil and gas are trapped in these holes. One of the most common sedimentary rocks is fine-grained shale. Layers of shale in the earth's crust four miles thick are common.

That is, shale is the majority of the rock in said layers. As time passed in the earth's history these layers of shale were deposited in the oceans. Continents were flooded over with water for eons and huge mounds of sediment were formed, which contain large amounts of shale and sandstone. But what happened to the oil and gas deposits? Many microscopic sea creatures—trillions of them have fallen—plant and animal alike and they became a speck each of the shale sediment on the ocean floor. Of course, larger sea animals like fish, sharks and whales have died contributed to this deposit, but their portion is small compared to the tiny plants and animals contribution—only because there have been trillions and trillions of them.

In the beginning oil and gas deposits are really small, but as the earth's humongous pressure squeezed these deposits they traveled horizontally and they run into other roaming oil. Eventually these moving spots of gas and oil are trapped in a rock formation of sandstone or limestone. Their roaming days are over. They can't detach from these sponge-like rocks. The two elementary types of traps are the structural trap and the strategic trap. A structural trap occurs when the earth has been bent or deformed. These structures may be in the form of a dome or a crease in the earth. It could also be caused by a more complex action like a vertical movement of the earthen layers. There are three types of structural trap: the salt dome trap, the anticline and the fault. Stratigraphic Traps are formed when sandstone gets closed up in shale.

The shale stops the oil and gas from roaming away. Otherwise, they would move when the earth exerted enough pressure on them again. Stratigraphic Gaps are missing divisions in the standard geographic column of the earth's sedimentary layers. These gaps are common. There is a problem in determining age or length of time involved in these gaps. Layers found in the surrounding area are missing in the gap. Those areas are dated with characteristic fossil types and radiometric dating. Many times there are stratigraphic gaps in the fossil record. Actually they are not the exception, but they are the norm.

There is a new view that says that these gaps don't represent an incomplete ness of the fossil record. In this view the stratigraphic data should be evaluated to determine if a phylogenetic relationship exists. Whereas, traditionally this relationship has been assumed, therefore it has been thought that the fossil record is incomplete.

GOLD DEPOSIT

Nevada is known for striking it rich in casinos and other gambling endeavors, but what many people do not know is that there is gold in them thar' hills. The great basin of Nevada is one of the world's greatest resources of stratigraphic gold deposits in the world. Almost 11% of the total world production of gold and gold deposits comes from this region. More than 3200 metric tons of gold is produced annually which equals out to 100,000,000 ounces of gold. This accounts for more than 10% of the world's total gold production.

A stratigraphic gold deposit can't keep a mining company or mining venture in business for many years. The importance of the stratigraphic gold deposit in the great basin area has a profound impact on our nation's economy and is the backbone of our world trade programs. The importance of this deposit is so ingrained in the economic success of our country that only state-of-the-art machinery and mining techniques are used to pull the gold from the ground. No longer are precious metals simply bulldozed or chopped out of the ground by shovel, picks, or axes. Delicate machinery gently pulls the gold from the earth and leaves little or no trace of environmental damage or ecological disarray. The United

States geological survey has become an active participant in the mining operations of the great basin.

They have coupled their successes and powers with the world of academia to produce top-of-the-line mining methods and also provide the groundwork for research and development in the future. The Nevada Bureau of mines and geology has also joined in with the mining industry and the United States geological survey. Together some of the greatest minds and methods have been brought together to protect this viable United States resource. The great basin project focused on questions about the environmental and ecological impact of gold mining and silver mining in the area. There is a world-class sedimentary rock that contains gold and silver composites found along the Carlin trend.

In the great basin there are also sedimentary exhalative deposits that are mined locally and have unique qualities and sulfurous concentration due to their procurement from sulfurous hydrothermal systems. These elaborate mining areas have posed some serious technological questions when it comes to mining methods and different environments. The accumulation of the academic and governmental forces seeks to answer these questions to check the integrity of the gold deposits and at the same time protect the integrity of the desert that is around them. The main objective of this endeavor is to find out the source in the evolutionary paths of hydrothermal fluids.

This is essential in improving the descriptive and genetic models of the great basin type deposits that are site-specific to stratigraphic gold deposits. Reconnaissance studies, mapping, and GPS location maps are being created to get the men and materials quickly to the sites and at the same time be able to protect the environment around the sites. Many other disciplines are combining forces to tackle global problems and the mining industry is leading the field in creating these types of partnerships.

GEOPHYSICS IN MINING

Jewelry is used by people of all ages and categories. These ornaments are made of different types of materials. Out of the various kinds of metals used to make jewelry, gold has a unique value. Gold jewelry has a special elegance in its look and appearance which increases the beauty of this yellow metal. With the increase in the value of gold, the value of our jewelry too increases.

The increased price of Gold has never affected the jewelry market. People continue to consider gold ornaments as a part of their pride and prestige and wish to have them as their asset. To meet the growing demands of gold, they are mined in several parts of the world. Gold for making jewelry is mined from earth and processed to make beautiful ornaments.

Mining of Gold requires a proper monitoring of the earth and its conditions before it is done. When mining increases beyond a limit, chances of affecting the existence of life on earth is high. So a separate branch of science must deal with all aspects of gold mining. It should properly check for the acquisition of data and processing and interpretation of collected data before proceeding with wide scale gold mining. Use of Geophysics in Mining for Gold used for Jewelry is high. They are used to explore the underground of earth and to get an image of the underground. Examples include: Perlen and Brautschmuck It can be seen that such a geophysical exploration is being used in various parts of the world where gold mining is done to help jewelry industry. Different tools and techniques of geophysics are used in mining gold.

When gold mining is performed on a large scale, various advanced and useful methods and techniques are to be used. This can be done with the help of experts who are skilled in analyzing the geophysical methods. The techniques of geophysics are used to find out the spots in earth where large quantities of gold are present. This will help mining to be conducted easily to meet the growing demands of gold these days. The gold obtained after mining is suitably processed to make fine metal. This is later used to design jewelry of different types. So the importance of geophysics in jewelry making is not negligible.

Understanding the importance of geophysics in mining of gold, various researches are going on in this area. This research is aimed at inventing advanced techniques for making mining simpler and easier. Environmental aspects of gold mining on large scale are also studied as a part of this research and study. When such studies are continued in various parts of the world, gold mining will turn out to be simple and easy and will be sufficient to meet the demands of the customers. When the importance of gold in jewelry making continues to be increasing, the chances of environmental exploitation also increase. A timely alert gains relevance. This has also increased the importance of geophysics in mining of gold. This hopes to increase the supply of gold in proper terms with the earthly conditions.

LATERITE IN COMPARISON TO GEOPHYSICS

The term laterite is becoming an issue in the world of geophysics. People who argue against the term claim that laterite is interchangeable with other words that mean a soil substrate or substance. Other proponents state that the term laterite is a borrowed definition that is taken from another scientific source that does not have to do with the discipline. There seems to be no clear definition though some state that it refers to the identification of the soil additives that make it more defined. Defenders of the term claim gives justice to its use in geophysics language though laterite is a very large class and cannot be grouped in any one homogenous class.

The idea that laterite is a single term that cannot define a single entity or substance has not been substantiated. Even when considered in a clay or hard soil sample, there are two many mineral and chemical compounds that makes that sample unique. The term is used more when the soil referred to has large amounts of aluminum compounds or iron. In the garden industry the term laterite refers to a soil that would have poor growth success because of the makeup of the soil and the number of metallic compounds found in it. This would mean that the metallic compounds were at such a level that the soil and dirt itself could not provide the organic material in sufficient amounts to grow new life. This is true of most all soils that are considered laterite.

The soils are never pure and contain materials that could be considered impurities such as sand, gravel, rocks, and organic materials. The materials are inert and non-reactive. This composition may be good to use in an aquatic environment that would not saturate itself with water or harm the organic resources for plant and animal growth. This would backfire if the laterite contained large amounts of iron. Iron would stimulate growth or the organic materials but in an aquatic or semi-damp environment invite the growth of algae, mosses, and other water borne parasitic and symbiotic growth.

Laterite is used in determining long term changes in atmosphere and weather by several different disciplines. The iron and clay mixture can take on qualities on the climate that they experience and can be used in finding out what that ancient climate was like in terms of moisture, dryness, cold, or hot. Once water has left the saturated laterite, fossil records can be studied to find out what life forms the era produced by matching the laterite with the flora or fauna fossilized within. This can be translated in the world of geology, geophysics, and other earth based sciences. There are many different types of laterite.

Though this article discussed iron, there are also nickel, aluminum, and other metallic components that can be infused in the clay or soil to be labeled as a laterite in geophysical terms. As the disciplines converge on this one term that means a different thing to each discipline, an understanding of the true meaning of laterite will arise and sub-categories of laterite may be used to explain the majority of the substance that makes the compound.

GROUND PENETRATING RADAR

With the advent of technology, locating the objects below the surface of the earth is made simple and there are various methods of identifying the underground water content. One method is by sending out impulses to the underground using devices. When these impulses reach the surface of underground objects, they will reflect the impulses back to the surface of the earth.

This is absorbed by the objects and this will then be analyzed with the help of various measurements and the presence is ensured. One of the devices working in this manner is known as Ground penetrating radars or GDR. Ground penetrating radars are used in geophysics to detect the presence of objects in the subsurface with the help of radar pulses.

This works with the help of microwaves of radio spectrum and uses the signals reflected from the underground object to detect the location. This finds a great number of applications in earth sciences, archeology, environmental remediation etc. this can be used in all categories of media including rock, soil, fresh water etc. to detect any change in the structure and presence of voids, cracks etc.

This method is used in the identification of pools present in the underground below earth's surface. These radars can easily identify the location of water tables and can identify the location of underground water contents without the need to excavate by going to the underground.

The most important aspect of a study in this way is that there are no harmful radiations or other dangerous effects when the presence of water is identified in this way with the help of GDR. They send in radar impulses, which when reaches the underground pool will be reflected back and this is put to further analysis and thus the location of water below the surface is identified.

A thorough analysis with the help of GDR will help to evaluate the quality of water in the underground pool. Since water might be contaminated due to the occurrence of various chemicals, a proper examination of this quality is needed to ensure that no damage or hazard is expected with the presence of chemicals in the underground pool water.

GDR can easily recognize the difference in density of materials even if they are present in the underground. They will be able to detect the presence of water in soil and this property is made use of in the detection of underground pools. There are some limitations to the device up to which it can identify the underground objects.

This limitation never hinders the performance of GDR in any way. The penetrating power of GDR is used in detecting the location of underground water contents. This is not easy with any other devices of similar kind. Contractors and other engineers are making use of this technique in identifying the presence of underground pools. In many environmental research and related areas, in the study of water and soil, this is the main method by which the presence of water below the surface of earth is understood.

GEOPHYSICS IN THE SEARCH FOR OIL

In the ancient times, oil was mainly used for medicinal purposes. In those days, oil was mainly collected from the surface. When the demand for oil increased, the traditional methods of finding oil were proved to be inefficient. It was at this time, that man started depending on geophysics to trace out the location of oil beneath the earth's surface. We are using reservoirs for collecting oil and hence it is hard to locate the hydro carbon rocks. Researchers are expected to look for hundreds of meters beneath the earth's surface.

This will be possible only with a proper support of tools with the assistance of geophysics. The explorers working in the area of geologists use various tools and accessories including map, log book and acids to locate the presence of oil. They observe and find out the possibility of hydro carbons in the underground. Once they search and find this out, they record them. They use different shots to get an idea of the chances of oil present in particular locations. The aerial shot and satellite shots are taken with the help of advanced techniques in geophysics. These photographs are combined with their observations.

This will result in the formation of a hypothesis. This hypothesis will support the fact that there are chances of oil presence down there and we can further look down to locate it. Now comes the turn of geophysics. Geophysicist examines the properties of sub soil. They use various methods to compare the results with that of the findings given by geologists and this is detailed in the Geology Dictionary. Gravimetry is used to measure gravity and to get an idea about the nature of the strata based on the density. The magnetic field is measured with magnetometry. If crystalline terrains are located, there is no chance of oil presence there.

A signal is emitted by the vibrator truck which will be reflected back and received by the geophones. This data is later transmitted to the truck in the laboratory. The sound waves generated by the surface shock will be refracted and reflected from the underground. The seismic recordings of the geophysics technology are later fed into the computers. Isochronic lines are used to mark the terrains. We will get images of the underground strata and this can be used to check if the strata contain hydrocarbons with the help of seismic waves. Ships are used for taking seismic measurements from the sea.

Once the data is processed, we will get an image which can be used for oil detection, thus leading to earlier planning for oil pipelines. Most of the data for this purpose is collected offshore. Seismic vessels and hydrophones make this task easier. The collective effort of a team of geophysics is important in searching for oil. These studies and research generally come up with a possibility of location of oil. In order to confirm the occurrence, a drilling is to be performed. Since this is an expensive process and have higher impacts, a drilling should not be performed without obtaining the conformation from the department of geophysics. This is the relevance of geophysics in connection with other techniques in the search of oil below the earth's surface.

ATLANTIC WARMER THAN A HOT TUB

The scientists of geophysics have always studied about earth in a quantitative manner. They have recently found out that the temperature of Atlantic Ocean has reached around 42 degree C which is about 14 deg C higher than the temperatures of ocean and it is warmer than a hot tub. This surprisingly high temperature in the oceans is the highest on Earth and occurred millions of years ago due to the high carbon dioxide level of the atmosphere of earth. The researchers are expecting this to go even higher in future than that is anticipated due to the heating of oceans by green house gases.

The result of the study shows the underestimation of future warming by the climate models. The Paleoclimatologist, Karen Bice has observed that the temperature reported in Atlantic Ocean is off the chart from what have noticed before. Their team of scientists researched on the sediment collected from the seafloor. The study on the sediments showed that they consisted of organic matter rich in carbon and had unusually high amounts of this matter accumulated in it. They also consisted of the shells of the fossilized organisms which have accumulated on the surface of the sea floor from millions of years ago. They studied on the isotopic nature of the shells and traced the chemistry of the elements.

These features are supposed to change with the change in the temperature on the surface of the water which was their main ecosystem. The temperature of the oceans at the time when dinosaurs dominated the Earth was found out to be between 33 deg C and 42 deg C. They have found out that the temperatures in the same region is around 28 deg C now akin to the rising temperature of a hottub. The same organic matter was put to study to find out the amount of carbon dioxide in the atmosphere. This created a situation where the scientists were put to tension on how this variation in carbon dioxide will increase the climate on earth's environment. They have also reported that if the previous studies were provided correct and accurate, the expectations of increase in temperature in ocean in the future will be even higher.

The studies also point to the change of the hydrologic cycle in earth. Winter storms and hurricanes are expected to increase in intensity due to temperature rise. They have observed from the studies that the temperature in the warm periods in olden times have released methane gas to the atmosphere and this gas is capable of absorbing temperature to a higher extent than carbon dioxide. This is missed out by the models. The high hottubs -like temperature in the Atlantic Ocean found out 91 million years ago were not inferred from this model of high methane content in the atmosphere. This means that the model is underestimating the response of the system to climatic conditions. The research team consisted of Bice and Kristina Dahl. Work of Bice was supported by Woods Hole Oceanographic Institution with their private funding. Funding was provided with the help of Joint Oceanographic Institutions U.S Science Support Programme.

THERMAL INSULATION IN GEOPHYSICS

Use of Thermal Insulation in Geophysics helps the user to understand a scientific concept or carry out difficult arithmetic operations carried during research. Use of Thermal Insulation in Geophysics Science application is a spreadsheet application used to search for scientific data and values, which may be also plotted in terms of a graph or pie chart. It was then extended to compare spectra of a test sample with standard spectrum images to able to distinguish one element or compound. Use of Thermal Insulation in Geophysics is used in numerous ways which include teaching a concept, or enabling quick arithmetic operations. The most effective way to help understand a concept is the simulation of the concept.

The simulation in the most basic sense can be an image representation, i.e. a virtually created image, for example an image of the molecular structure of a given compound. To make an elaborate representation would mean to create a video clip of the same. Example would be to show the use of Thermal Insulation in Geophysics and its reaction occurring over the time of the reaction. It may also include a simplified database application used to look up values or properties corresponding to a given input. These looked up values may be inputs to other application themselves or maybe just used directly by the user themselves. The most important type of study on the use of Thermal Insulation in Geophysics is emulations.

Given, that scientist are unsure of the result and possible extreme situation when a hypothesized project would overreact, causing heavy monetary and in certain severe research project loss of life itself, emulations are used. Emulations of such projects provide a basic idea such as Einblasdämmung - Wärmedämmung, and estimation which can be optimized toward the actual implementation of that project. In cases where loss of life is possible like training students on medical equipment or controlling an airplane or even construction of a skyscraper, emulations play an inevitable role.

It would be insane to actually let an inexperienced person to operate either a medical device or fly an airplane filled with passengers. Emulations are used first, in which students undergo training in which; even a critical situation is simulated over the emulation, so as to enable the interns to handle a similar real life case. The same is true in case of the Use of Thermal Insulation in Geophysics The Use of Thermal Insulation in Geophysics furthermore include those software that would extrapolate scientific data, and also would be able to tell what the affects would have happened in the environment under research had the various variable been different. To enable the user to have ample data, apart from entered data to understand, manipulate and look into to make a fair and through conclusion.

What about the Use of Thermal Insulation in Geophysics which uses scientific knowledge as a base. It is the implementation purely on the basic of scientific understanding. In order to understand how to Use of Thermal Insulation in Geophysics, detailed study and observations are considered important, especially of past attempts and records.

Chapter 2

Soil Mechanics

Soil mechanics is a discipline that applies principles of engineering mechanics, e.g. kinematics, dynamics, fluid mechanics, and mechanics of material, to predict the mechanical behaviour of soils. Together with Rock mechanics, it is the basis for solving many engineering problems in civil engineering (geotechnical engineering), geophysical engineering and engineering geology.

Some of the basic theories of soil mechanics are the basic description and classification of soil, effective stress, shear strength, consolidation, lateral earth pressure, bearing capacity, slope stability, and permeability. Foundations, embankments, retaining walls, earthworks and underground openings are all designed in part with theories from soil mechanics.

BASIC CHARACTERISTICS OF SOILS

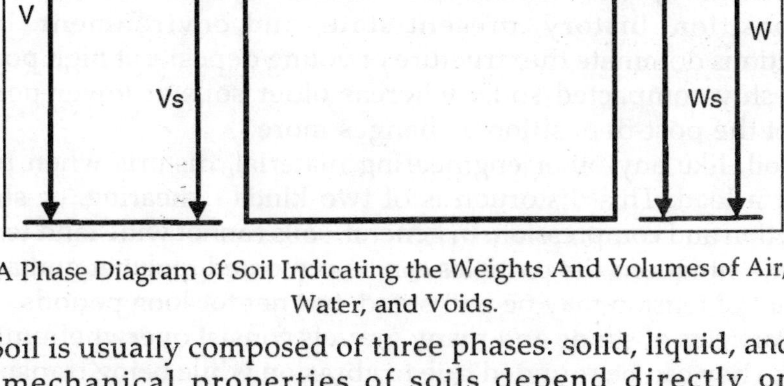

Fig. A Phase Diagram of Soil Indicating the Weights And Volumes of Air, Soil, Water, and Voids.

Soil is usually composed of three phases: solid, liquid, and gas. The mechanical properties of soils depend directly on the interactions of these phases with each other and with applied potentials (e.g., stress, hydraulic head, electrical potential, and temperature difference).

The solid phase of soils contains various amounts of crystalline clay and non-clay minerals, noncrystalline clay material, organic matter, and precipitated salts. These minerals are commonly formed by atoms of elements such as oxygen, silicon, hydrogen, and aluminum, organized in various crystalline forms. These elements along with calcium, sodium, potassium, magnesium, and carbon comprise over 99% of the solid mass of soils. Although, the amount of non-clay material is greater than that of clay and organic material, the latter have a greater influence in the behaviour of soils. Solid particles are classified by size as clay, silt, sand, gravel, cobbles, or boulders.

The liquid phase in soils is commonly composed of water containing various types and amounts of dissolved electrolytes. Organic compounds, both soluble and immiscible are present in soils from chemical spills, leaking wastes, and contaminated groundwater.

The gas phase, in partially saturated soils, is usually air, although organic gases may be present in zones of high biological activity or in chemically contaminated soils.

Soil mineralogy controls the size, shape, and physical and chemical properties of soil particles and thus its load-carrying ability and compressibility.

The *structure* of a soil is the combined effects of *fabric* (particle association, geometrical arrangement of particles, particle groups, and pore spaces in a soil), *composition*, and interparticle forces. The structure of soils is also used to account for differences between the properties of natural (structured) and remolded soils (destructured). The structure of a soil reflects all facets of the soil composition, history, present state, and environment. Initial conditions dominate the structure of young deposits at high porosity or freshly compacted soils; whereas older soils at lower porosity reflect the post-depositional changes more.

Soil, like any other engineering material, distorts when placed under a load. This distortion is of two kinds - shearing, or sliding, distortion and compression. In general, soils cannot withstand tension. In some situations the particles can be cemented together and a small amount of tension may be withstood, but not for long periods.

Particles of sands and many gravels consist overwhelmingly of silica. They can be rounded due to abrasion while being transported by wind or water, or sharp-cornered, or anything in between, and are roughly equi-dimensional. Clay particles arise from weathering of rock crystals like feldspar, and commonly consist of alumino-silicate minerals. They generally have a flake-shape with a large surface area compared with their mass. As their mass is extremely small, their behaviour is governed by forces of electrostatic attraction and repulsion on their surfaces. These forces attract and adsorb water to their surfaces, with the thickness of the layer being affected by dissolved salts in the water.

Sieve Analysis

Sieve analysis is the process of determining the size of soil particles by passing the soil sample through a number of different sieves having different opening (hole sizes).

Sieve #	Length of one side of opening (mm)
4	4.75
10	2.0
20	0.850
40	0.425

EFFECTIVE STRESS

Karl von Terzaghi first proposed the relationship for effective stress in 1936. For him, the term 'effective' meant the calculated stress that was effective in moving soil, or causing displacements. It represents the average stress carried by the soil skeleton.

Effective stress (σ') acting on a soil is calculated from two parameters, total stress (σ) and pore water pressure (u) according to:

$$\sigma' = \sigma - u$$

Typically, for simple examples

$$\sigma = H_{soil\gamma soil}$$

$$u = H_{w\gamma w}$$

Much like the concept of stress itself, the formula is a construct, for the easier visualization of forces acting on a soil mass, especially simple analysis models for slope stability, involving a slip plane. With these models, it is important to know the total weight of the soil above (including water), and the pore water pressure within the slip plane, assuming it is acting as a confined layer.

However, the formula becomes confusing when considering the true behaviour of the soil particles under different measurable conditions, since none of the parameters are actually independent actors on the particles.

Fig. Arrangement of Spheres Showing Contacts

Consider a grouping of round quartz sand grains, piled loosely, in a classic 'cannonball' arrangement. As can be seen, there is a contact stress where the spheres actually touch. Pile on more spheres and the contact stresses increase, to the point of causing frictional instability (dynamic friction), and perhaps failure.

The independent parameter affecting the contacts (both normal and shear) is the force of the spheres above. This can be calculated by using the overall average density of the spheres and the height of spheres above.

Fig. Spheres Immersed in Water, Reducing Effective Stress

If we then have these spheres in a beaker and add some water, they will begin to float a little depending on their density (buoyancy). With natural soil materials, the effect can be significant, as anyone who has lifted a large rock out of a lake can attest.

The contact stress on the spheres decreases as the beaker is filled to the top of the spheres, but then nothing changes if more water is added. Although the water pressure between the spheres (pore water pressure) is increasing, the effective stress remains the same, because the concept of 'total stress' includes the weight of all the water above. This is where the equation can become confusing, and the effective stress can be calculated using the buoyant density of the spheres (soil), and the height of the soil above.

Fig. Spheres Being Injected with Water, Reducing Effective Stress

The concept of effective stress truly becomes interesting when dealing with non-hydrostatic pore water pressure. Under the conditions of a pore pressure gr .dient, the ground water flows, according to the permeability equation (Darcy's law). Using our spheres as a model, this is the same as injecting (or withdrawing) water between the spheres. If water is being injected, the seepage force acts to separate the spheres and reduces the effective stress. Thus, the soil mass becomes weaker. If water is being withdrawn, the spheres are forced together and the effective stress increases.

Two extremes of this effect are quicksand, where the groundwater gradient and seepage force act against gravity; and the 'sandcastle effect', where the water drainage and capillary action act to strengthen the sand. As well, effective stress plays an important role in slope stability, and other geotechnical engineering and engineering geology problems, such as groundwater-related subsidence.

SOIL SHEAR STRENGTH

Shear strength in reference to soil is a term used to describe the maximum strength of soil at which point significant plastic deformation or yielding occurs due to an applied shear stress. There is no definitive "shear strength" of a soil as it depends on a number of factors affecting the soil at any given time and on the frame of reference, in particular the rate at which the shearing occurs.

Two theories are commonly used to estimate the shear strength of a soil depending on the rate of shearing as a frame of reference. These are Tresca theory for short term loading of a soil, commonly referred to as the undrained strength or the total stress condition and Mohr–Coulomb theory combined with the principle of effective stress for the long term loading of a soil, commonly referred to as the drained strength or the effective stress condition.

In modern soil mechanics, both these classical approaches (Tresca and Mohr–Coulomb) may be superseded by critical state theory or by steady state theory either of which can be considered in both undrained and drained terms and also cases involving partial drainage. The classical approaches are still in common usage; however, both are taught in undergraduate civil engineering programs, and consequently, they are also used in practice.

Shear strength of a soil is also of importance in designing for earthquakes where the concept of the soil's steady state shear strength is used.

The shear strength of soils is taught in detail in specialist masters degree programs. Such programs usually include the use of modern numerical modeling techniques such as finite element analysis coupled with a model for shear strength such as critical state soil mechanics.

Factors Controlling Shear Strength of Soils

The stress-strain relationship of soils, and therefore the shearing strength, is affected by:

- *Soil composition (basic soil material):* mineralogy, grain size and grain size distribution, shape of particles, pore fluid type and content, ions on grain and in pore fluid.
- *State (initial):* Defined by the initial void ratio, effective normal stress and shear stress (stress history). State can be described by terms such as: loose, dense, overconsolidated, normally consolidated, stiff, soft, contractive, dilative, etc.
- *Structure:* Refers to the arrangement of particles within the soil mass; the manner the particles are packed or distributed. Features such as layers, joints, fissures, slickensides, voids, pockets, cementation, etc, are part of the structure. Structure of soils is described by terms such as: undisturbed, disturbed, remolded, compacted, cemented; flocculent, honey-combed, single-grained; flocculated, deflocculated; stratified, layered, laminated; isotropic and anisotropic.
- *Loading conditions:* Effective stress path, i.e., drained, and undrained; and type of loading, i.e., magnitude, rate (static, dynamic), and time history (monotonic, cyclic)).

Undrained Strength

This term describes a type of shear strength in soil mechanics as distinct from drained strength.

Conceptually, there is no such thing as the *undrained strength of a soil. It depends on a number of factors, the main ones being:*

- Orientation of stresses
- Stress path
- Rate of shearing
- Volume of material (like for fissured clays or rock mass)

Undrained strength is typically defined by Tresca theory, based on Mohr's circle as:

$$\sigma_1 - \sigma_3 = 2\,S_u$$

Where:

σ_1 is the major principal stress
σ_3 is the minor principal stress
τ is the shear strength $(\sigma_1 - \sigma_3)/2$
hence, $\tau = S_u$ (or sometimes c_u), the undrained strength.

It is commonly adopted in limit equilibrium analyses where the rate of loading is very much greater than the rate at which pore water pressures, that are generated due to the action of shearing the soil, may dissipate. An example of this is rapid loading of sands during an earthquake, or the failure of a clay slope during heavy rain, and applies to most failures that occur during construction.

As an implication of undrained condition, no elastic volumetric strains occur, and thus Poisson's ratio is assumed to remain 0.5 throughout shearing. The Tresca soil model also assumes no plastic volumetric strains occur. This is of significance in more advanced analyses such as in finite element analysis. In these advanced analysis methods, soil models other than Tresca may be used to model the undrained condition including Mohr-Coulomb and critical state soil models such as the modified Cam-clay model, provided Poisson's ratio is maintained at 0.5.

One relationship used extensively by practicing engineers is the empirical SHANSEP (stress history and normalized soil engineering properties) relationship. This is based on the observation that the logarithm of the undrained shear strength S_u normalized by the vertical consolidation stress σ_{vc} plots linearly against the logarithm of the over consolidation ratio or OCR, that is, $S_u/\sigma_{vc}=K^*OCR^N$ where K and N are constants that depend on the soil and the loading used to shear the soil. This relationship can be derived from critical state soil mechanics.

Drained Strength

This term describes a type of shear strength in soil mechanics as distinct from undrained strength.

The drained strength is the strength of the soil when pore water pressures, generated during the course of shearing the soil, are able to dissipate rapidly. It also applies where no pore water exists in the soil (the soil is dry). It is commonly defined using Mohr-Coulomb theory (it was called "Coulomb's equation" by Karl von Terzaghi in 1942 combined with the principle of effective stress.

Drained strength is defined as:

$\tau = \sigma' \tan(\varphi') + c'$

Where $\sigma' = (\sigma - u)$, *known as the principle of effective stress.* σ *is the total stress applied normal to the shear plane, and* u *is the pore water pressure acting on the same plane.*

φ' = the effective angle of shearing resistance. Formerly termed 'angle of internal friction' after Coulomb friction, where the coefficient of friction ì is equal to tan(ϕ), which is proportional to the normal force on a plane but independent of its area. It is now regarded to have little to do with friction, and more to do with the micro-mechanical interaction of soil particles. It has sometimes been referred to as the "angle of repose" as a dry granular material will form a pile at this angle but no steeper. It is further described as either peak ϕ'_p, critical state ϕ'_{cv} or residual ϕ'_r. Note that ϕ'_p is only adopted in relation to Terzaghi's misunderstanding of the nature of "true" cohesion. Nowadays, critical state ϕ'_{cv} values should be prescribed.

c' = apparent cohesion. Allows the soil to possess some shear strength at no confining stress, or even under tensile stress. Commonly ascribed to temporary negative pore water pressures (suction), that dissipate over time. It may also be due to diagenetic affects caused by soil aging such as chemical bonding, cementation of grains and the effects of creep; indeed Coulomb identified that soil possessed no cohesion when newly remoulded, Heyman 1972 as these diagenetic effects had been destroyed. When shear tests are conducted on an overconsolidated or dense soil, and peak strengths are plotted on a τ/σ plot, it appears that cohesion exists as the y-intercept is non-zero. Some feel that this is not due to true cohesion, but is the effect of interlocking of particles.

In any case, the long term loading condition must rely on the soil properties expected to exist and contribute to the shear strength of the soil over the long term, and for these reasons it is generally not considered a reliable soil mechanical property unlike φ'.

CRITICAL STATE SOIL MECHANICS

Critical State Soil Mechanics is the area of Soil Mechanics that encompasses the conceptual models that represent the mechanical behaviour of saturated remolded soils based on the *Critical State* concept. The Critical State concept is an idealization of the observed behaviour of saturated remoulded clays in triaxial compression tests, and it is assumed to apply to undisturbed soils. It states that soils and other granular materials, if continuously distorted (sheared) until they flow as a frictional fluid, will come into a well-defined critical state. At the onset of the critical state, shear distortions ε_s occur without any further changes in mean effective stress, p' deviatoric stress q (or yield stress, σ_y, in uniaxial tension according to the von Mises yielding criterion), or specific volume v:

$$\frac{\partial p'}{\partial \varepsilon_s} = \frac{\partial q}{\partial \varepsilon_s} = \frac{\partial v}{\partial \varepsilon_s} = 0$$

where,

$$v = 1 + e$$

$$P' = \frac{1}{3}(\sigma'_1 + \sigma'_2 + \sigma'_3)$$

$$q = \sqrt{\frac{(\sigma'_1 - \sigma'_2) + (\sigma'_2 - \sigma'_3) + (\sigma'_1 - \sigma'_3)^2}{3}}$$

However, for triaxial conditions $\sigma'_2 = \sigma'_3$. Thus,

$$P' = \frac{1}{3}(\sigma'_1 + 2\sigma'_3)$$

$$q = (\sigma'_2 = \sigma'_3)$$

All critical states, for a given soil, form a unique line called the *Critical State Line* (*CSL*) defined by the following equations in the space (p', q, u):

$$q = Mp'$$

$$v = \Gamma - \lambda \ln(p')$$

where, M, Γ, and are soil constants. The first equation determines the magnitude of the deviatoric stress needed to keep the soil flowing continuously as the product of a frictional constant M(capital μ) and the mean effective stress p'. The second equation states that the specific volume occupied by unit volume of flowing particles will decrease as the logarithm of the mean effective stress increases.

A more advanced understanding of the behaviour of soil undergoing shearing lead to the development of the critical state theory of soil mechanics. In critical state soil mechanics, a distinct shear strength is identified where the soil undergoing shear does so at a constant volume, also called the 'critical state'. Thus there are three commonly identified shear strengths for a soil undergoing shear:

- Peak strength τ_p
- Critical state or constant volume strength τ_{cv}
- Residual strength τ_r

The peak strength may occur before or at critical state, depending on the initial state of the soil particles being sheared:

- A loose soil will contract in volume on shearing, and may not develop any peak strength above critical state. In this case 'peak' strength will coincide with the critical state shear strength, once the soil has ceased contracting in volume. It may be stated that such soils do not exhibit a distinct 'peak strength'.
- A dense soil may contract slightly before granular interlock prevents further contraction (granular interlock is dependent on the shape of the grains and their initial packing arrangement). In order to continue shearing once granular interlock has occurred, the soil must dilate (expand in volume). As additional shear force is required to dilate the soil, a 'peak' strength occurs. Once this peak strength caused by dilation has been overcome through continued shearing, the resistance provided by the soil to the applied shear stress reduces (termed "strain softening"). Strain softening will continue until no further changes in volume of the soil occur on continued shearing. Peak strengths are also observed in overconsolidated clays where the natural fabric of the soil must be destroyed prior to reaching constant volume shearing. Other effects that result in peak strengths include cementation and bonding of particles.

The constant volume (or critical state) shear strength is said to be intrinsic to the soil, and independent of the initial density or packing arrangement of the soil grains. In this state the grains being sheared are said to be 'tumbling' over one another, with no significant granular interlock or sliding plane development affecting the resistance to shearing. At this point, no inherited fabric or bonding of the soil grains affects the soil strength.

The residual strength occurs for some soils where the shape of the particles that make up the soil become aligned during shearing (forming a slickenside), resulting in reduced resistance to continued shearing (further strain softening). This is particularly true for most clays that comprise plate-like minerals, but is also observed in some granular soils with more elongate shaped grains. Clays that do not have plate-like minerals (like allophanic clays) do not tend to exhibit residual strengths.

Use in practice: If one is to adopt critical state theory and take c′ = 0; τ_p may be used, provided the level of anticipated strains are taken into account, and the effects of potential rupture or strain softening to critical state strengths are considered. For large strain deformation, the potential to form slickensided surface with a ϕ'_r should be considered (such as pile driving).

The Critical State occurs at the quasi-static strain rate. It does not allow for differences in shear strength based on different strain rates. Also at the critical state, there is no particle alignment or specific soil structure.

SOIL CONSOLIDATION

Consolidation is a process by which soils decrease in volume. It occurs when stress is applied to a soil that causes the soil particles to pack together more tightly, therefore reducing its bulk volume. When this occurs in a soil that is saturated with water, water will be squeezed out of the soil. The magnitude of consolidation can be predicted by many different methods. In the Classical Method, developed by Karl von Terzaghi, soils are tested with an oedometer test to determine their compression index. This can be used to predict the amount of consolidation.

When stress is removed from a consolidated soil, the soil will rebound, regaining some of the volume it had lost in the consolidation process. If the stress is reapplied, the soil will consolidate again along a recompression curve, defined by the recompression index. The soil which had its load removed is considered to be *overconsolidated*. This is the case for soils which have previously had glaciers on them. The highest stress that it has been subjected to is termed the *preconsolidation stress*. The *over consolidation ratio* or OCR is defined as the highest stress experienced divided by the current stress. A soil which is currently experiencing its highest stress is said to be *normally consolidated* and to have an OCR of one. A soil could be considered *underconsolidated* immediately after a new load is applied but before the excess pore water pressure has had time to dissipate.

CONSOLIDATION ANALYSIS

Spring Analogy

The process of consolidation is often explained with an idealized system composed of a spring, a container with a hole in its cover, and water. In this system, the spring represents the compressibility or the structure itself of the soil, and the water which fills the container represents the pore water in the soil.

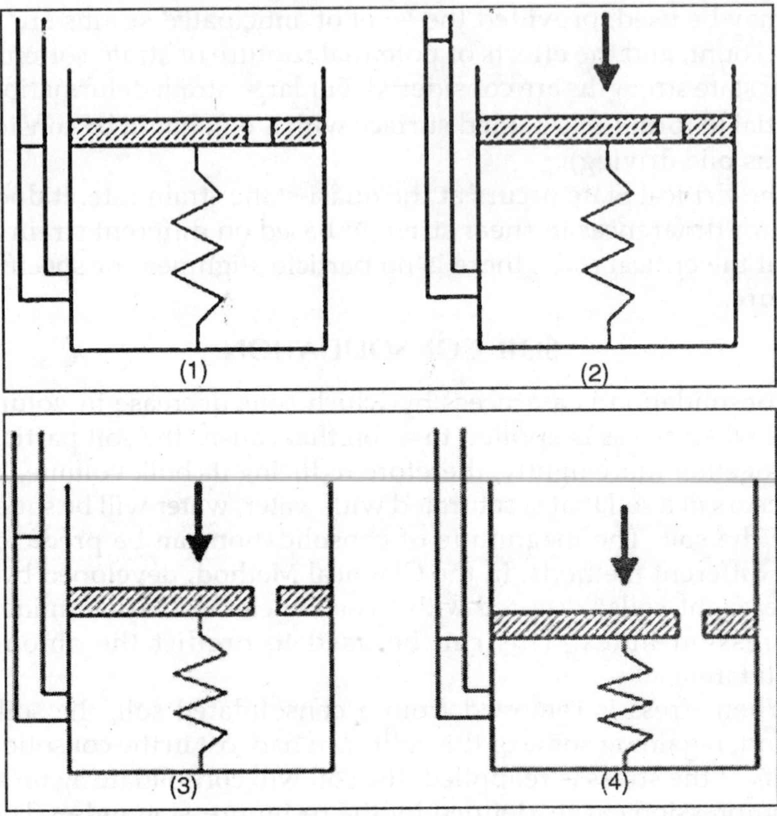

- The container is completely filled with water, and the hole is closed. (Fully saturated soil)
- A load is applied onto the cover, while the hole is still unopened. At this stage, only the water resists the applied load. (Development of excessive pore water pressure)
- As soon as the hole is opened, water starts to drain out through the hole and the spring shortens. (Drainage of excessive pore water)
- After some time, the drainage of water no longer occurs. Now, the spring alone resists the applied load. (Full dissipation of excessive pore water pressure. End of consolidation)

Primary Consolidation

This method assumes consolidation occurs in only one-dimension. Laboratory data is used to construct a plot of strain or void ratio verses effective stress where the effective stress axis is on a logarithmic scale. The plot's slope is the compression index or recompession index. The equation for consolidation settlement of a normally consolidated soil can then be determined to be:

$$\delta_c = \frac{C_c}{1+e_o} H \log\left(\frac{\sigma' \approx_f}{\sigma' \approx_o}\right)$$

where

δ_c is the settlement due to consolidation.

C_c is the compression index.

e_0 is the initial void ratio.

H is the height of the soil.

σ_{zf} is the final vertical stress.

σ_{z0} is the initial vertical stress.

C_c can be replaced by C_r (the recompession index) for use in overconsolidated soils where the final effective stress is less than the preconsolidation stress. When the final effective stress is greater than the preconsolidation stress, the two equations must be used in combination to model both the recompression portion and the virgin compression portion of the consolidation process, as follows:

$$\delta_c = \frac{C_c}{1+e_o} H \log\left(\frac{\sigma'_{zc}}{\sigma'_{zo}}\right) + \frac{C_c}{1+e_o} H \log\left(\frac{\sigma'_{zf}}{\sigma'_{zc}}\right)$$

where σ_{zc} is the preconsolidation stress of the soil.

Secondary Consolidation

Secondary consolidation is the compression of soil that takes place after primary consolidation. Secondary consolidation is caused by creep, viscous behaviour of the clay-water system, compression of organic matter, and other processes. In sand, settlement caused by secondary compression is negligible, but in peat, it is very significant.

Secondary consolidation is given by the formula

Where H_0 is the height of the consolidating medium

e_0 is the initial void ratio

C_a is the secondary compression index

Time Dependency

The time for consolidation to occur can be predicted. Sometimes consolidation can take years. This is especially true in saturated clays because their hydraulic conductivity is extremely low, and this causes the water to take an exceptionally long time to drain out of the soil.

While drainage is occurring, the pore water pressure is greater than normal because it is carrying part of the applied stress (as opposed to the soil particles).

LATERAL EARTH PRESSURE

Lateral earth pressure is the pressure that soil exerts in the horizontal plane. The common applications of lateral earth pressure theory are for the design of ground engineering structures such as retaining walls, basements, tunnels, and to determine the friction on the sides of deep foundations.

To describe the pressure a soil will exert, a lateral earth pressure coefficient, K, is used. K is the ratio of lateral (horizontal) pressure to vertical pressure ($K = \sigma_h'/\sigma_v'$). Thus horizontal earth pressure is assumed to be directly proportional to the vertical pressure at any given point in the soil profile. K can depend on the soil properties and the stress history of the soil. Lateral earth pressure coefficients are broken up into three categories: at-rest, active, and passive.

The pressure coefficient used in geotechnical engineering analyses depends on the characteristics of its application. There are many theories for predicting lateral earth pressure; some are empirically based, and some are analytically derived.

At rest Pressure

At rest lateral earth pressure, represented as K_0, is the in situ horizontal pressure. It can be measured directly by a dilatometer test (DMT) or a borehole pressuremeter test (PMT). As these are rather expensive tests, empirical relations have been created in order to predict at rest pressure with less involved soil testing, and relate to the angle of shearing resistance. Two of the more commonly used are presented below.

Jaky for normally consolidated soils:

$$K_{o(NC)} = 1 - \sin\phi'$$

Mayne & Kulhawy for overconsolidated soils:

$$K_{o(OC)} = K_{O(NC)} * OCR^{(\sin\phi')}$$

The latter requires the OCR profile with depth to be determined. To estimate K_0 due to compaction pressures.

Active and Passive Pressure

The active state occurs when a soil mass is allowed to relax or move outward to the point of reaching the limiting strength of the soil; that is, the soil is at the failure condition in extension. Thus it is the minimum lateral soil pressure that may be exerted. Conversely, the passive state occurs when a soil mass is externally forced to the limiting strength (that is, failure) of the soil in compression. It is the maximum lateral soil pressure that may be exerted. Thus active and passive pressures define the minimum and maximum possible pressures respectively that may be exerted in a horizontal plane.

Simplified Explanation of Typical Retaining Walls

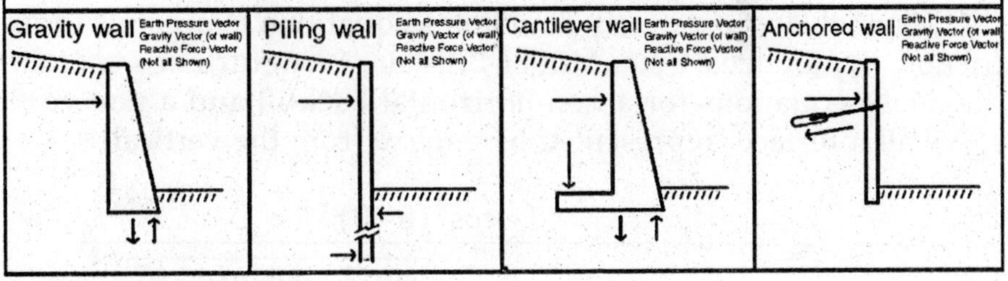

Rankine Theory

Rankine's theory, developed in 1857' is a stress field solution that predicts active and passive earth pressure. It assumes that the soil is cohesionless, the wall is frictionless, the soil-wall interface is vertical, the failure surface on which the soil moves is planar, and the resultant force is angled parallel to the backfill surface. The equations for active and passive lateral earth pressure coefficients are given below. Note that ϕ' is the angle of shearing resistance of the soil and the backfill is inclined at angle β to the horizontal.

$$K_a = \cos\beta \frac{\cos\beta - \left(\cos^2\beta - \cos^2\phi\right)^{1/2}}{\cos\beta + \left(\cos^2\beta - \cos^2\phi\right)^{1/2}}$$

$$K_P = \cos\beta \frac{\cos\beta + \left(\cos^2\beta - \cos^2\phi\right)^{1/2}}{\cos\beta - \left(\cos^2\beta - \cos^2\phi\right)^{1/2}}$$

For the case where β is 0, the above equations simplify to

$$K_a = \tan^2\left(45 - \frac{\phi}{2}\right)$$

$$K_P = \tan^2\left(45 + \frac{\phi}{2}\right)$$

Coulomb Theory

Coulomb first studied the problem of lateral earth pressures on retaining structures. He used limit equilibrium theory, which considers the failing soil block as a free body in order to determine the limiting horizontal earth pressure. The limiting horizontal pressures at failure in extension or compression are used to determine the K_a and K_p respectively. Since the problem is indeterminate a number of potential failure surfaces must be analysed to identify the critical failure surface (i.e. the surface that produces the maximum or minimum thrust on the wall).

Mayniel later extended Coulomb's equations to account for wall friction, symbolized by *ä*. Müller-Breslau (1906) further generalized Mayniel's equations for a non-horizontal backfill and a non-vertical soil-wall interface (represented by angle è from the vertical).

$$K_a = \frac{\cos^2(\phi-\theta)}{\cos^2\theta\cos(\delta+\theta)\left(1+\sqrt{\frac{\sin(\delta+\phi)\sin(\phi-\beta)}{\cos(\delta+\theta)\cos(\beta-\theta)}}\right)^2}$$

$$K_P = \frac{\cos^2(\phi-\theta)}{\cos^2\theta\cos(\delta-\theta)\left(1+\sqrt{\frac{\sin(\delta+\phi)\sin(\phi-\beta)}{\cos(\delta+\theta)\cos(\beta-\theta)}}\right)^2}$$

Caquot and Kerisel

In 1948, Albert Caquot (1881-1976) and Jean Kerisel (1908-2005) developed an advanced theory that modified Muller-Breslau's equations to account for a non-planar rupture surface. They used a logarithmic spiral to represent the rupture surface instead. This modification is extremely important for passive earth pressure where there is soil-wall friction. Mayniel and Muller-Breslau's equations are unconservative in this situation and are dangerous to apply. For the active pressure coefficient, the logarithmic spiral rupture surface provides a negligible difference compared to Muller-Breslau. These equations are too complex to use, so tables or computers are used instead..

Equivalent fluid Pressure

Terzaghi and Peck, in 1948, developed empirical charts for predicting lateral pressures. Only the soil's classification and backfill slope angle are necessary to use the charts. In order for this method to be safe to implement, the values given are conservative. This method was developed with Rankine theory.

Bell's Relation

For soils with cohesion, Bell developed an analytical solution that uses the square root of the pressure coefficient to predict the cohesion's contribution to the overall resulting pressure. These equations represent the total lateral earth pressure. The first term represents the non-cohesive contribution and the second term the cohesive contribution. The first equation is for an active situation and the second for passive situations.

$$\sigma_h = K_a\sigma_u - 2c\sqrt{K_a}$$

$$\sigma_h = K_a\sigma_u + 2c\sqrt{K_a}$$

BEARING CAPACITY

In geotechnical engineering, bearing capacity is the capacity of soil to support the loads applied to the ground. The bearing capacity of soil is the maximum average contact pressure between the foundation and the soil which should not produce shear failure in the soil. Ultimate bearing capacity is the theoretical maximum pressure which can be supported without failure; *while allowable bearing capacity is the ultimate bearing capacity divided by a factor of safety.*

Sometimes, on soft soil sites, large settlements may occur under loaded foundations without actual shear failure occurring; in such cases, the allowable bearing capacity is based on the maximum allowable settlement.

There are three modes of failure that limit bearing capacity: general shear failure, local shear failure, and punching shear failure.

General Shear Failure

The general shear failure case is the one normally analyzed. Prevention against other failure modes is accounted for implicitly in settlement calculations. There are many different methods for computing when this failure will occur.

Terzaghi

Karl von Terzaghi developed a method for determining bearing capacity for the general shear failure case in 1943. The equations are given below.

For square foundations:

$$qult = 1.3'N_c + \sigma'_{zD}N_q + 0.4\gamma'BN_\gamma$$

For continuous foundations:

$$qult = c'N_c + \sigma'_{zD}N_q + 0.5\gamma'BN_\gamma$$

For circular foundations:

$$qult = 1.3c'N_c + \sigma'_{zD}N_q + 0.3\gamma'BN_\gamma$$

where

$$N_q = \frac{e^{2\pi(0.75-\phi/360)\tan\phi'}}{2\cos^2(45+\phi'/2)}$$

$$N_c = \frac{Nq-1}{\tan\phi'}\ \text{for}\varphi'>$$

$$N\gamma = \frac{\tan\phi'}{2}\left(\frac{K_{p\gamma}}{\cos^2\phi'}-1\right)$$

*c*2 is the effective cohesion.

σ_{zD}2 is the vertical effective stress at the depth the foundation is laid.

γ 2 is the effective unit weight when saturated or the total unit weight when not fully saturated.

B is the width or the diameter of the foundation.

φ 2 is the effective internal angle of friction.

$K_{p\gamma}$ is obtained graphically. Simplifications have been made to eliminate the need for $K_{p\gamma}$. One such was done by Coduto, given below, and it is accurate to within 10%.

$$N\gamma = \frac{2(Nq+1)\tan\phi'}{1+0.4\sin 4\phi'}$$

SLOPE STABILITY

The field of slope stability encompasses the analysis of static and dynamic stability of slopes of earth and rock-fill dams, slopes of other types of embankments, excavated slopes, and natural slopes in soil and soft rock.

As seen in Figure, earthen slopes can develop a cut-spherical weakness zone. The probability of this happening can be calculated in advance using a simple 2-D circular analysis package. A primary difficulty with analysis is locating the most-probable slip plane for any given situation. Many landslides have only been analyzed after the fact.

It should be noted that real life failures in naturally deposited mixed soils are not necessarily circular, but prior to computers, it was far easier to analyse such a simplified geometry. Nevertheless, failures in 'pure' clay can be quite close to circular. Such slips often occur after a period of heavy rain, when the pore water pressure at the slip surface increases, reducing the effective normal stress and thus diminishing the restraining friction along the slip line. This is combined with increased soil weight due to the added groundwater. A 'shrinkage' crack (formed during prior dry weather) at the top of the slip may also fill with rain water, pushing the slip forward. At the other extreme, slab-shaped slips on hillsides can remove a layer of soil from the top of the underlying bedrock. Again, this is usually initiated by heavy rain, sometimes combined with increased loading from new buildings or removal of support at the toe (resulting from road widening or other construction work). Stability can thus be significantly improved by installing drainage paths to reduce the destabilising forces. Once the slip has occurred, however, a weakness along the slip circle remains, which may then recur at the next monsoon.

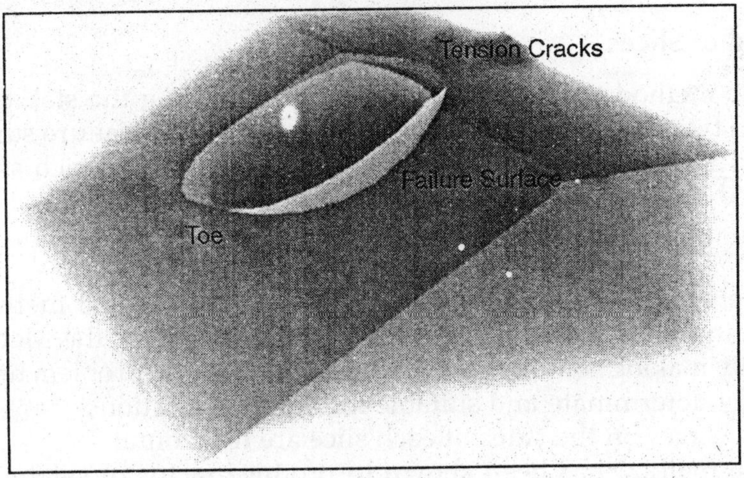

Fig. Simple Slope Slip Section

Slope stability issues can be seen with almost any walk down a ravine in an urban setting. If the toe is eroded too far, or the swimming pool begins to leak, the forces driving a slope failure will exceed those resisting failure, and a landslide will develop, possibly quite suddenly.

Analysis Methods

If the forces available to resist movement are greater than the forces driving movement, the slope is considered stable. A factor of safety is calculated by dividing the forces resisting movement by the forces driving movement. In earthquake-prone areas, the analysis is typically run for static conditions and pseudo-static conditions, where the seismic forces from an earthquake are assumed to add static loads to the analysis.

Fig. Slope with Eroding River and Swimming Pool

Method of Slices

The method of slices is a method for analyzing the stability of a slope in two dimensions. The sliding mass above the failure surface is divided into a number of slices. The forces acting on each slice are obtained by considering the mechanical equilibrium for the slices.

Bishop's Method

The Modified (or Simplified) Bishop's Method is a method for calculating the stability of slopes. It is an extension of the Method of Slices. By making some simplifying assumptions, the problem becomes statically determinate and suitable for hand calculations:

- Forces on the sides of each slice are horizontal

The method has been shown to produce factor of safety values within a few per cent of the "correct" values.

$$F = \frac{\sum \left[\frac{c' + \left((W/b) - u \right) \tan \phi'}{\Psi} \right]}{\sum \left[(W/b) \sin \alpha \right]}$$

where

$$\Psi = \cos \alpha + \frac{\sin \alpha \tan \phi}{F}$$

c' is the effective cohesion

ϕ' is the effective internal angle of friction

b is the width of each slice, assuming that all slices have the same width

W is the weight of each slice

u is the water pressure at the base of each slice

Permeability and Seepage

Permeability in the earth sciences (commonly symbolized as ê, or k) is a measure of the ability of a material (typically, a rock or unconsolidated material) to transmit fluids. It is of great importance in determining the flow characteristics of hydrocarbons in oil and gas reservoirs, and of groundwater in aquifers. It is typically measured in the lab by application of Darcy's law under steady state conditions or, more generally, by application of various solutions to the diffusion equation for unsteady flow conditions.

Formula

The intrinsic permeability of any porous material is:

$$\kappa_I = C \cdot d^2$$

where

κ_I is the intrinsic permeability [L^2]

C is a dimensionless constant that is related to the configuration of the flow-paths

d is the average, or effective pore diameter [L]

Permeability needs to be measured, either directly (using Darcy's law) or through estimation using empirically derived formulas.

A common unit for permeability is the *darcy* (D), or more commonly the *millidarcy* (mD) (1 darcy $\approx 10^{-12} m^2$). Other units are cm^2 and the SI m^2.

Permeability is part of the proportionality constant in Darcy's law which relates discharge (flow rate) and fluid physical properties (e.g. viscosity), to a pressure gradient applied to the porous media. The proportionality constant specifically for the flow of water through a porous media is the hydraulic conductivity; permeability is a portion of this, and is a property of the porous media only, not the fluid. In naturally occurring materials, it ranges over many orders of magnitude.

For a rock to be considered as an exploitable hydrocarbon reservoir without stimulation, its permeability must be greater than approximately 100 mD (depending on the nature of the hydrocarbon - gas reservoirs with lower permeabilities are still exploitable because of the lower viscosity of gas with respect to oil). Rocks with permeabilities significantly lower than 100 mD can form efficient *seals*. Unconsolidated sands may have permeabilities of over 5000 mD.

Tensor Permeability

To model permeability in anisotropic media, a permeability tensor is needed. Pressure can be applied in three directions, and for each direction, permeability can be measured (via Darcy's law in 3D) in three directions, thus leading to a 3 by 3 tensor. The tensor is realized using a 3 by 3 matrix being both symmetric and positive definite (SPD matrix):

- The tensor is symmetric by the Onsager reciprocal relations.
- The tensor is positive definite as the component of the flow parallel to the pressure drop is always in the same direction as the pressure drop.

The permeability tensor is always diagonalizable (being both symmetric and positive definite). The eigenvectors will yield the principal directions of flow, meaning the directions where flow is parallel to the pressure drop, and the eigenvalues representing the principal permeabilities.

Chapter 3

Rotation and Revolution

The term Earth rotation refers to the spinning of our planet on its axis. Because of rotation, the Earth's surface moves at the equator at a speed of about 467 m per second or slightly over 1675 km per hour. If we could look down at the Earth's North Pole from space we would notice that the direction of rotation is counter-clockwise. The opposite is true if the Earth is viewed from the South Pole. One rotation takes exactly twenty-four hours and is called a mean solar day. The Earth's rotation is responsible for the daily cycles of day and night. At any one moment in time, one half of the Earth is in sunlight, while the other half is in darkness. The edge dividing the daylight from night is called the circle of illumination. The Earth's rotation also creates the apparent movement of the Sun across the horizon.

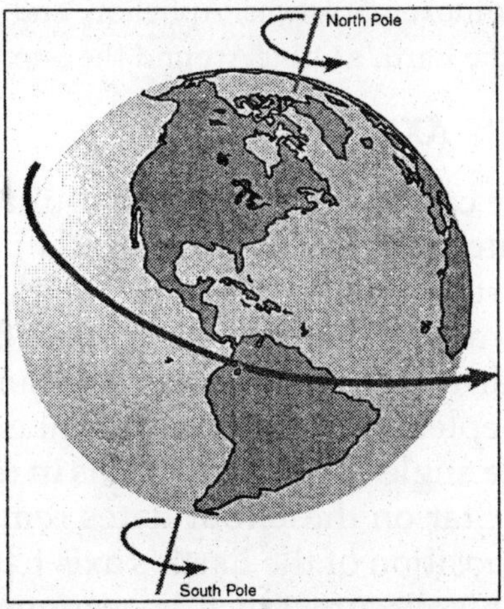

Fig. The Movement of the Earth about its Axis is known as Rotation.

The orbit of the Earth around the Sun is called an Earth revolution. This celestial motion takes 365.26 days to complete one cycle. Further, the Earth's orbit around the Sun is not circular, but oval or elliptical. An elliptical orbit causes the Earth's distance from the Sun to vary over a year. Yet, this phenomenon is not responsible for the Earth's seasons!

This variation in the distance from the Sun causes the amount of solar radiation received by the Earth to annually vary by about 6%. Figure illustrates the positions in the Earth's revolution where it is closest and farthest from the Sun. On January 3, perihelion, the Earth is closest to the Sun (147.3 million km). The Earth is farthest from the Sun on July 4, or aphelion (152.1 million km). The average distance of the Earth from the Sun over a one-year period is about 149.6 million km.

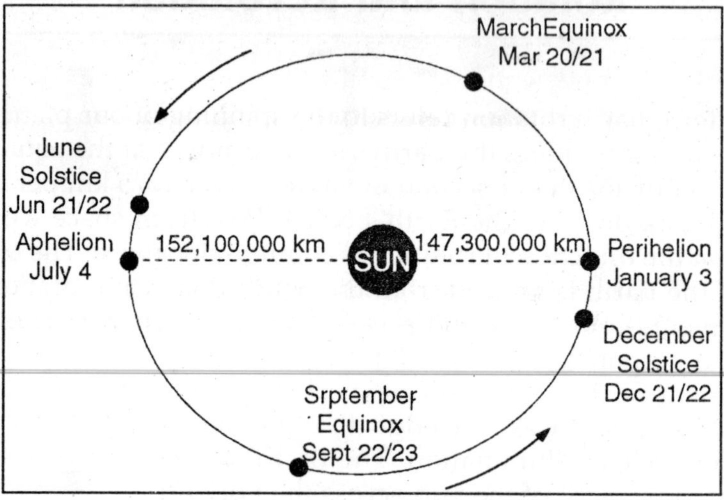

Fig. Position of the Equinoxes, Solstices, Aphelion, and Perihelion Relative to the Earth's Orbit Around the Sun.

TILT OF THE EARTH'S AXIS

The ecliptic plane can be defined as a two-dimensional flat surface that geometrically intersects the Earth's orbital path around the Sun. On this plane, the Earth's axis is not at right angles to this surface, but inclined at an angle of about 23.5° from the perpendicular. Figure shows a side view of the Earth in its orbit about the Sun on four important dates: June solstice, September equinox, December solstice, and March equinox. Note that the angle of the Earth's axis in relation to the ecliptic plane and the North Star on these four dates remains unchanged.

Yet, the relative position of the Earth's axis to the Sun does change during this cycle. This circumstance is responsible for the annual changes in the height of the Sun above the horizon. It also causes the seasons, by controlling the intensity and duration of sunlight received by locations on the Earth. Figure shows an overhead view of this same phenomenon. In this view, we can see how the circle of illumination changes its position on the Earth's surface. During the two equinoxes, the circle of illumination cuts through the North Pole and the South

Pole. On the June solstice, the circle of illumination is tangent to the Arctic Circle (66.5° N) and the region above this latitude receives 24 hours of daylight. The Arctic Circle is in 24 hours of darkness during the December solstice.

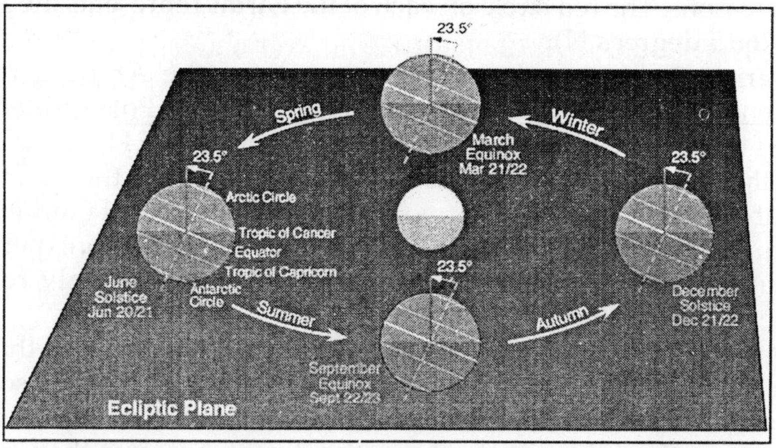

Fig. The Earth's Rotational Axis is Tilted 23.5° from the Red line Drawn Perpendicular to the Ecliptic plane. This tilt remains the same anywhere along the Earth's orbit around the Sun. Seasons are appropriate only for the Northern Hemisphere.

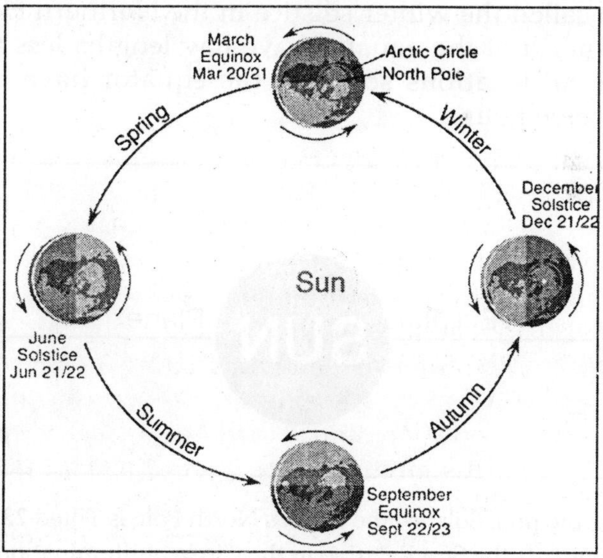

Fig. Annual Change in the Position of the Earth in its Revolution Around the Sun.

In this graphic, we are viewing the Earth from a position in space that is above the North Pole (yellow dot) at the summer solstice, the winter solstice, and the two equinoxes. Note how the position of the North Pole on the Earth's surface does not change. However, its position relative to the Sun does change and this shift is responsible for the seasons. The red circle on each of the Earths represents the Arctic Circle (66.5 degrees N).

During the June solstice, the area above the Arctic Circle is experiencing 24 hours of daylight because the North Pole is tilted 23.5 degrees toward the Sun. The Arctic Circle experiences 24 hours of night when the North Pole is tilted 23.5 degrees away from the Sun in the December solstice. During the two equinoxes, the circle of illumination cuts through the polar axis and all locations on the Earth experience 12 hours of day and night. Seasons are appropriate only for the Northern Hemisphere.

On June 21 or 22 the Earth is positioned in its orbit so that the North Pole is leaning 23.5° toward the Sun. During the June solstice (also called the summer solstice in the Northern Hemisphere), all locations north of the equator have day lengths greater than twelve hours, while all locations south of the equator have day lengths less than twelve hours.

On December 21 or 22 the Earth is positioned so that the South Pole is leaning 23.5 degrees toward the Sun. During the December solstice (also called the winter solstice in the Northern Hemisphere), all locations north of the equator have day lengths less than twelve hours, while all locations south of the equator have day lengths exceeding twelve hours.

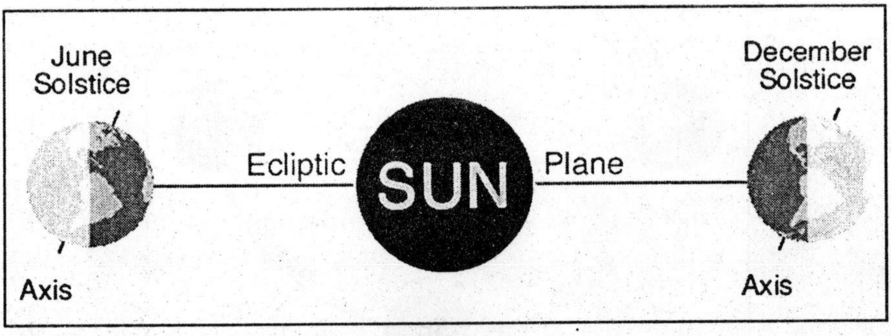

Fig. During the June Solstice the Earth's North Pole is Tilted 23.5 Degrees Towards the Sun Relative to the Circle of Illumination.

This phenomenon keeps all places above a latitude of 66.5 degrees N in 24 hours of sunlight, while locations below a latitude of 66.5 degrees S are in darkness. The North Pole is tilted 23.5 degrees away from the Sun relative to the circle of illumination during the December solstice. On this date, all places above a latitude of 66.5 degrees N are now in darkness, while locations below a latitude of 66.5 degrees S receive 24 hours of daylight.

On September 22 or 23, also called the autumnal equinox in the Northern Hemisphere, neither pole is tilted toward or away from the Sun. In the Northern Hemisphere, March 20 or 21 marks the arrival of the vernal equinox or spring when once again the poles are not tilted toward or away from the Sun. Day lengths on both of these days, regardless of latitude, are exactly 12 hours.

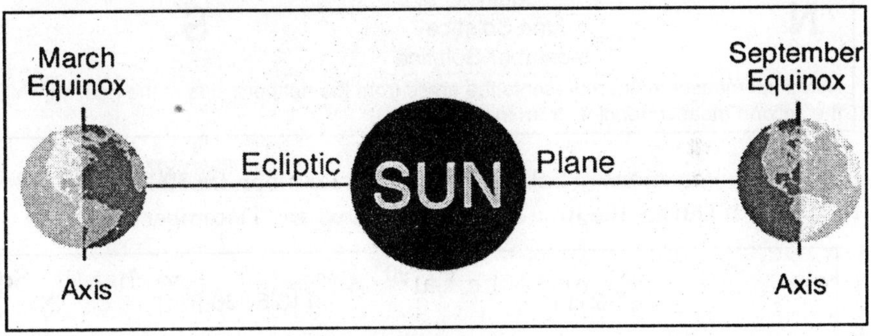

Fig. During the Equinoxes, the Axis of the Earth is not Tilted Toward or Away from the Sun and the Circle of Illumination cuts Through the Poles.

AXIS TILT AND SOLAR ALTITUDE

The annual change in the relative position of the Earth's axis in relationship to the Sun causes the height of the Sun or solar altitude to vary in our skies.

Solar altitude is normally measured from either the southern or northern point along the horizon and begins at zero degrees. Maximum solar altitude occurs when the Sun is directly overhead and has a value of 90°. The total variation in maximum solar altitude for any location on the Earth over a one-year period is 47° (Earth's tilt 23.5° x 2 = 47°).

This variation is due to the annual changes in the relative position of the Earth to the Sun. At 50 degrees North, maximum solar altitude varies from 63.5 degrees on the June solstice to 16.5 degrees on the December solstice. Maximum solar height at the equator goes from 66.5 degrees above the northern end of the horizon during the June solstice, to directly overhead on the September equinox, and then down to 66.5 degrees above the southern end of the horizon during the December solstice.

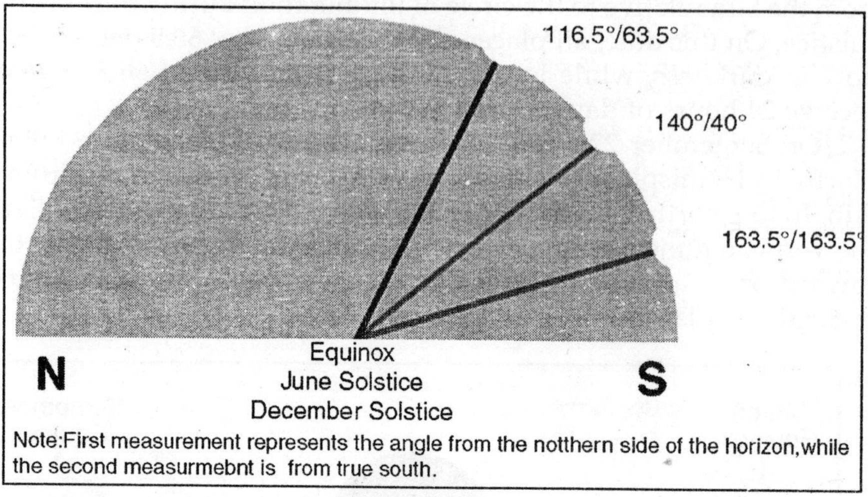

Fig. Variations in Solar Altitude at Solar Noon for 50 Degrees North During the June Solstice, Equinox, and December Solstice.

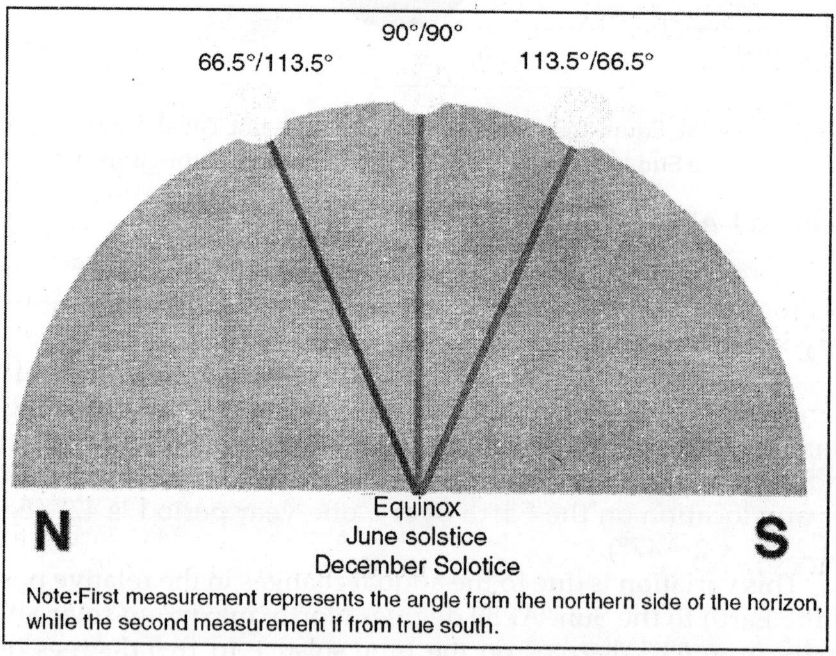

Fig. Variations in Solar Altitude at Solar Noon for the Equator During the June Solstice, Equinox, and December Solstice.

The location on the Earth where the Sun is directly overhead at solar noon is known as the subsolar point. The subsolar point occurs on the equator during the two equinoxes. On these dates, the equator is lined up with the ecliptic plane and the poles are in line with the circle of illumination. During the summer solstice, the subsolar point moves to the *Tropic of Cancer* (23.5° N) because at this time the North Pole is inclined 23.5° toward the Sun. Figure shows how the subsolar point gradually changes from one day to the next over a period of one-year. Note that on this graph, the subsolar point is located at the *Tropic of Capricorn* (23.5° S) during the December solstice when the South Pole is angled 23.5° toward the Sun.

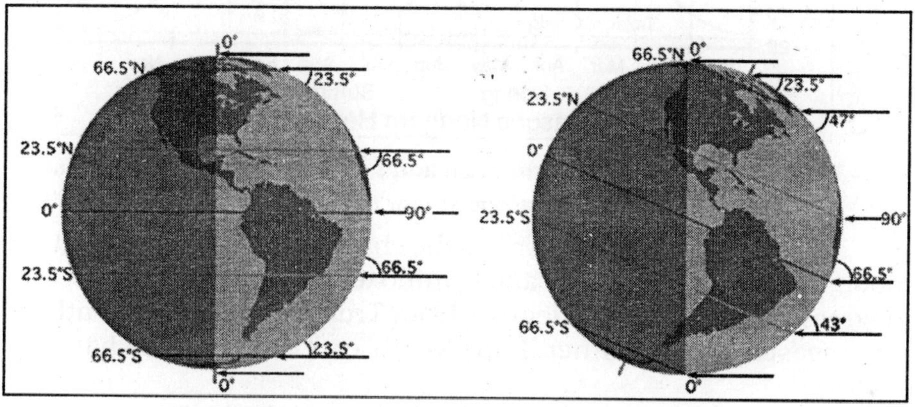

Fig. Relationship of Maximum Sun Height to Latitude for the Equinox (left) and June Solstice (right).

The red values on the right of the globes are maximum solar altitudes at solar noon. Black numbers on the left indicate the location of the Equator, Tropic of Cancer (23.5 degrees N), Tropic of Capricorn (23.5 degrees S), Arctic Circle (66.5 degrees N), and the Antarctic Circle (66.5 degrees S). The location of the North and South Poles are also identified. During the equinox, the equator is the location on the Earth with a Sun angle of 90 degrees for solar noon. Note how maximum Sun height declines with latitude as we move away from the Equator.

For each degree of latitude traveled maximum Sun height decreases by the same amount. At equinox, we can also calculate the noon angle by subtracting the location's latitude from 90. During the summer solstice, the Sun is now directly overhead at the Tropic of Cancer. All locations above this location have maximum Sun heights that are 23.5 degrees higher from the equinox situation. Places above the Arctic Circle are in 24 hours of daylight. Below the Tropic of Cancer the noon angle of the Sun drops one degree in height for each degree of latitude traveled. At the Antarctic Circle, maximum Sun height becomes 0 degrees and locations south of this point on the Earth are in 24 hours of darkness.

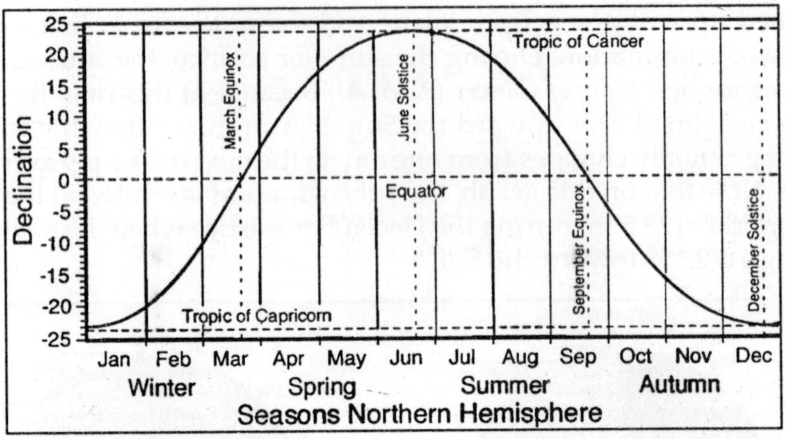

Fig. Angle of the Sun's Declination and Latitude of the Subsolar Point Throughout the year. Seasons are for the Northern Hemisphere.

The following table describes the changes in solar altitude at solar noon for the two solstices and equinoxes. All measurements are in degrees (horizon has 180 degrees from True North to True South) and are measured from either True North or True South (whatever is closer).

Table: Maximum Sun Altitudes for Selected Latitudes During the two Solstices and Equinoxes.

Location's Latitude	March Equinox March 20/21	June Solstice June 21/22	September Equinox September 22/23	December Solstice December 21/22
90 N	0 degrees	23.5 degrees	0 degrees	- 23.5 degrees
70 N	20 degrees	43.5 degrees	20 degrees	-3.5 degrees
66.5 N	23.5 degrees	47 degrees	23.5 degrees	0 degrees
60 N	30 degrees	53.5 degrees	30 degrees	6.5 degrees
50 N	40 degrees	63.5 degrees	40 degrees	16.5 degrees
23.5 N	66.5 degrees	90 degrees	66.5 degrees	43 degrees
0 degrees	90 degrees	66.5 degrees	90 degrees	66.5 degrees
23.5 S	66.5 degrees	43 degrees	66.5 degrees	90 degrees
50 S	40 degrees	16.5 degrees	40 degrees	63.5 degrees
60 S	30 degrees	6.5 degrees	30 degrees	53.5 degrees
66.5 S	23.5 degrees	0 degrees	23.5 degrees	47 degrees
70 S	20 degrees	-3.5 degrees	20 degrees	43.5 degrees
90 S	0 degrees	- 23.5 degrees	0 degrees	23.5 degrees

The following links show graphical illustrations of the annual movements of the Sun in our skies for selected latitudes. In these illustrations, solar angles are measured from both True North and True South for solar noon.

I 90 N I 66.5 N I 50 N I 23.5 N I Equator (0) I 23.5 S I 50 S I 66.5 S I 90 S I

Finally, the altitude of the Sun at solar noon can also be calculated with the following simple equation:

Altitude A = 90 - Latitude L +/- Declination D

In this is equation, L is the latitude of the location in degrees and D is the declination. The equation is simplified to A = 90 - L if Sun angle determinations are being made for either equinox date. If the Sun angle determination is for a solstice date, declination (D) is added to latitude (L) if the location is experiencing summer (northern latitudes = June solstice; southern latitudes = December solstice) and subtracted from latitude (L) if the location is experiencing winter (northern latitudes = December solstice; southern latitudes = June solstice).

All answers from this equation are given relative to True North for southern latitudes and True South for northern latitudes. For our purposes only the declinations of the two solstices and two equinoxes are important. These values are: June solstice D=23.5, December solstice D=23.5, March equinox D=0, and Septemeber equinox D=0. When using the above equation in tropical latitudes, Sun altitude values greater than 90 degrees may occur for some calculations. When this occurs, the noonday Sun is actually behind we when looking towards the equator. Under these circumstances, Sun altitude should be recalculated as follows:

Altitude A = 90 - (originally calculated Altitude A - 90)

EARTH-SUN RELATIONS

Solar radiation is one of many sources of energy, and probably one of the most important sources, that drive environmental processes acting at the surface of the Earth. The amount and intensity of solar radiation reaching the Earth is affected by the geometric relationship of the Earth with respect to the Sun.

THE SOLAR CONSTANT

Though the temperature of the air near the ground is primarily determined by the heat released by Earth's surface, the principal source for heating the Earth is solar radiation. The Earth is "constantly" bathed in solar radiation propagated through space ultimately reaching the surface of the earth. On average, the Earth receives 1368 W/m^2 (1.96 ly/min) of solar radiation at the outer edge of the atmosphere, called the "solar constant". However, the actual amount received at the edge of the atmosphere or the Earth's surface varies from place to place and day to day on account of the orientation of the Earth to the Sun.

EARTH REVOLUTION AND ROTATION

Earth, the third planet of our solar system revolves around the Sun once every 365 1/4 days. The elliptical orbit of the earth varies from 91.5 million miles on January 3 called "perihelion", to 94.5 million miles on July 4 called "aphelion" for an average earth-sun distance of 93 million miles. The elliptical path causes only small variations in the amount of solar radiation reaching the earth.

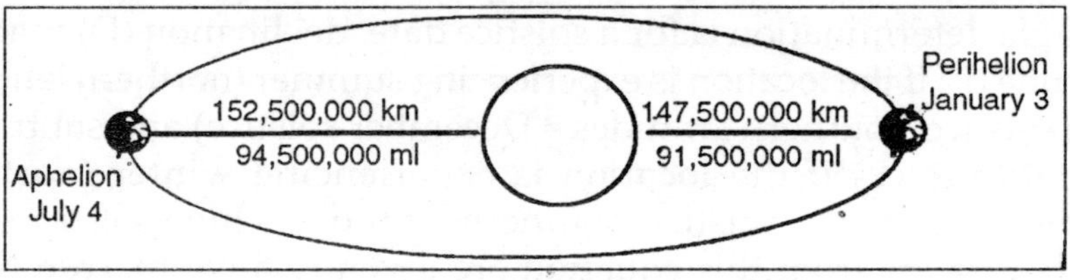

Fig. Earth's Elliptical Orbit

The Earth rotates at a uniform rate on its axis once every 24 hours. Turning in an eastward direction the Sun "rises" in the east and seemingly "travels" toward the west during the day. The Sun isn't actually moving, it's the eastward rotation towards the morning Sun that makes it appear that way. The Earth then rotates in the opposite direction to the apparent path of the Sun. Looking down from the North Pole yields a counterclockwise direction. From over the South Pole a clockwise direction of rotation occurs. We can demonstrate this by looking down at the North Pole of a counterclockwise rotating globe. Lift the globe while keeping it spinning in a counterclockwise direction and look at it from below.

Axial Tilt

The plane of the ecliptic is a plane that cuts through the centre of the Earth and the Sun in which the Earth revolves around the Sun. The Earth's axis is tilted 23 1/2 degrees from being perpendicular to the plane of the ecliptic. The axis of rotation remains pointing in the same direction as it revolves around the Sun, pointing toward the star Polaris. As a result, the Earth's axis of rotation remains parallel to its previous position as it orbits the sun, a property called "parallelism".

The constant tilt and parallelism causes changes in the angle that a beam of light makes with respect to a point on Earth during the year, called the "sun angle". The most intense incoming solar radiation occurs where the sun's rays strike the Earth at the highest angle. As the sun angle decreases, the beam of light is spread over a larger area and decreases in intensity. During the summer months the Earth is inclined toward the Sun yielding high sun angles. During the winter, the Earth is oriented away from the Sun creating low sun angles. The tilt of the Earth and its impact on sun angle is the reason the Northern and Southern Hemisphere have opposite seasons. Summer occurs when a hemisphere is tipped toward the Sun and winter when it is tipped away from the Sun.

Fig. Effect of Axial Tilt on Incoming Solar Radiation.

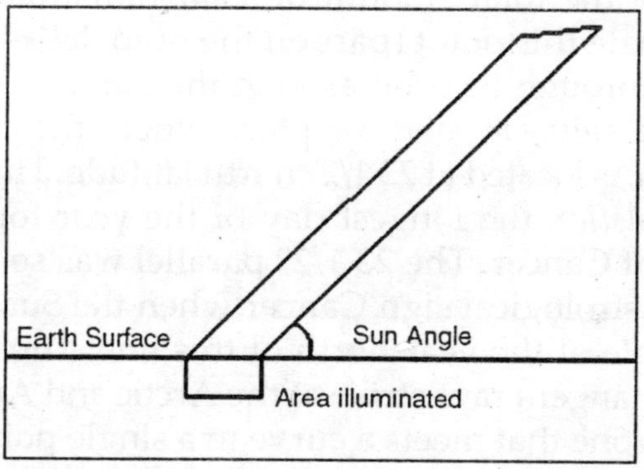

Fig. Sun Angle Determines the Intensity of Energy.

Path length and Insolation

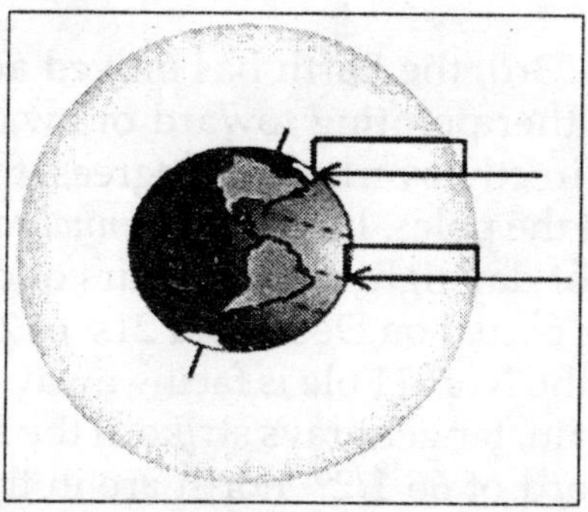

Fig. Influence of Path Length on Insolation

The distance that a beam of light travels greatly affects the amount of solar energy that ultimately reaches the Earth. The Earth-Sun distance only varies by about 3 million miles compared to an average distance of about 93 million miles over the year. A more significant impact on insolation is the thickness of the atmosphere on depletion of a beam of light. As the amount of atmosphere through which the beam passes increases, the greater the chance for reflection and scattering of light, thus reducing insolation at the surface. Due to the curvature of the Earth, a beam of light striking the Equator passes through less atmosphere than one at a higher latitude.

SEASONS AND EARTH-SUN RELATIONS

The Earth's axis always remains pointing in the same direction as it revolves around the sun. As a result, the solar angle varies at a given place throughout the year. The variation in sun angle is the prime cause of our seasons. The orientation of the Earth with respect to the Sun also determines the length of day. Together, the sun angle and day length determine the total amount of solar radiation incident at the Earth. To illuminate this point (pardon the pun), let's follow the Earth as it progresses through its orbit around the Sun.

At noon, the *subsolar point*, or place where the sun lies directly overhead at noon, is located at 23 1/2° north latitude. This date is known as the *summer solstice*, the longest day of the year for places located north of Tropic of Cancer. The 23 1/2° parallel was so named because it is during the astrological sign Cancer when the Sun's rays strike at their highest angle of the year north of this line. The North pole tips into the Sun and tangent rays strike at the Arctic and Antarctic Circles. (A tangent ray is one that meets a curve in a single point). This creates a 24 hour period of daylight ("polar day") for places located poleward of 66 1/2° north. We find the South Pole tipped away from the Sun, sending places poleward of 66 1/2° south into 24 hours of darkness ("polar night").

On Sept 22nd or 23rd, the Earth has moved around the Sun such that the poles are neither pointing toward or away from the sun. On this day, the Sun is directly overhead 0 degrees, the equator, at noon. Tangent rays strike at the poles. It is the *autumnal equinox* and all places experience 12 hours of day light and 12 hours of darkness.

The *winter solstice* occurs on December 21st or 22nd when the Earth has oriented itself so the North Pole is facing away from, and the South Pole into the Sun. Again, tangent rays strike at the Arctic and Antarctic circles. Places poleward of 66 1/2° north are in the grips of the cold, polar night. Places poleward of 66 1/2° south experience the 24 hour polar day. The Sun lies directly over 23 1/2° south. Occurring during the astrological sign of Capricorn, 23 1/2° south latitude is called the Tropic of Capricorn.

Continuing to March 20th or 21st (i.e. *Spring Equinox*) the Earth has positioned itself similar to that which occurs in September, only on the other side of the Sun. Once again tangent rays strike at the North and South poles, and the perpendicular rays of the Sun strike the Equator at noon. All places have equal day length (12 hours day;12 hours of night) as the circle of illumination cuts all latitudes in half.

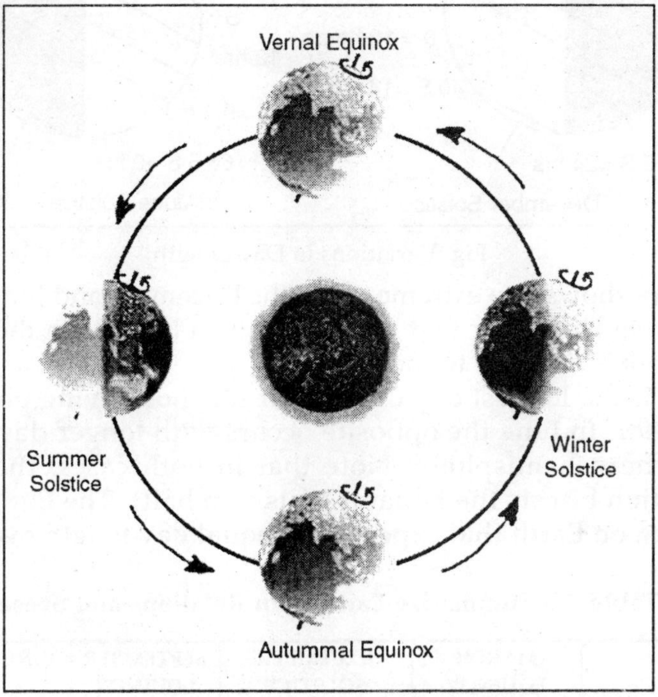

Fig. Orientation of Earth to Sun

So over the course of a year, the Sun's rays are only perpendicular to the surface (directly over head) at places between 23 1/2° north and south. Places between the Tropic of Cancer and Capricorn experience two times when the Sun is directly over head over the course of a year. The sun angle does not vary much for places between 23 1/2° north and south, a larger range in sun angle occurs poleward of these latitudes. The greater the variation in sun angle, the greater the variation in surface heating.

Day Length and Seasons

Day length is determined by the length of time the Sun is above the horizon. Day length changes through the year as the orientation of the Earth to the Sun changes. The circle of illumination is the imaginary circle that separate day from night.

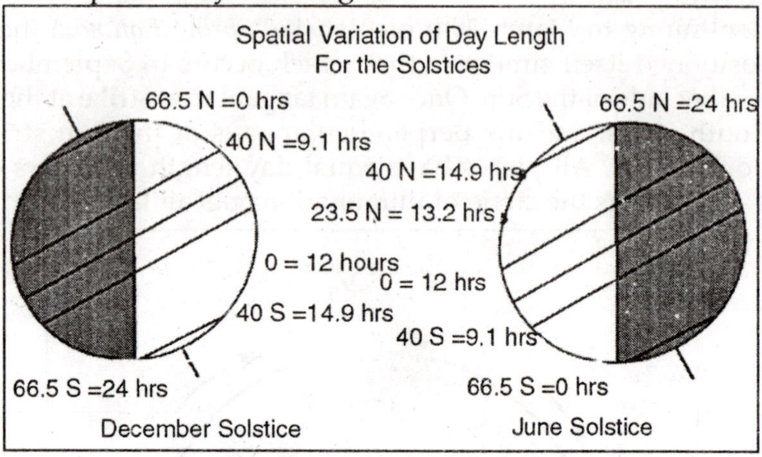

Fig. Variations in Day Length

Figure shows two extreme cases, the December and June solstices. Note during December that more of a given latitude in the Southern hemisphere is exposed to the Sun.

This is the longest day of the year for those living poleward of the Equator. In June the opposite occurs with longer day length in the Northern hemisphere. Note that in both cases, the circle of illumination bisects the Equator (cuts it in half). The Equator is the only place on Earth that experiences equal day length every day of the year.

Table. To Summarize Earth- Sun Relations and Season

	MARCH EQUINOX	DECEMBER SOLSTICE	SEPTEMBER EQUINOX	JUNE SOLSTICE
Date	March 21	December 22	Sept. 23	June 21
Subsolar Point	0°	23 1/2° S	0°	23 1/2° N
Tangent Rays	North and South Poles	Arctic and Antarctic Circles	North and South Poles	Arctic and Antarctic Circles
Day length	12 hour day length everywhere	24 hours of darkness at North Pole; 24 hours day light at South Pole; 12 hours day light at Equator	12 hour day length everywhere	24 hours of darkness at South Pole; 24 hours day light at North Pole; 12 hours day light at Equator

RADIATION

Most of the environmental processes acting near the surface of the

Earth derive their energy from exchanges of heat between the Earth and the atmosphere above. Much of this heat comes from radiant energy initially provided by the absorption of solar radiation. The absorbed energy is used to warm the atmosphere, evaporate water, warm the subsurface along with a host of other processes.

The Radiation Balance

The radiation balance of the Earth system is an accounting of the incoming and outgoing components of radiation. These components are balanced over long time periods and over the Earth as whole. If they weren't the Earth would be continually cooling or warming. However, over a short period of time, radiant energy is unequally distributed over the Earth.

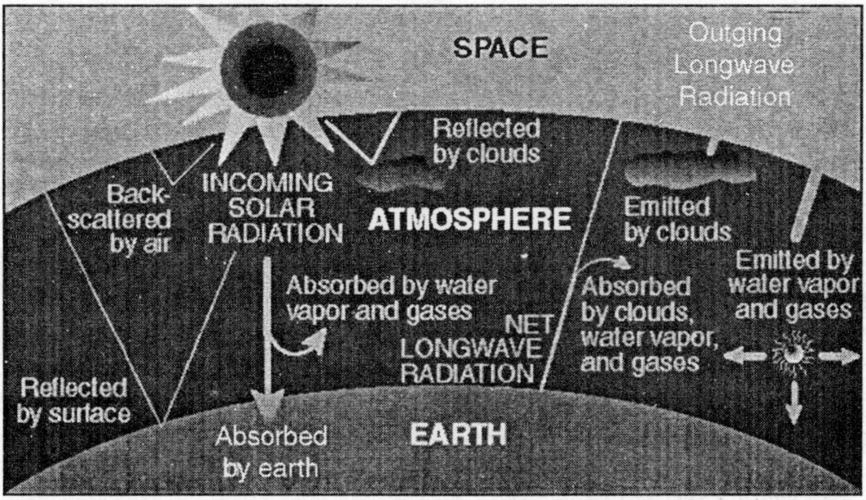

Fig. Radiation Balance of the Earth

Shortwave Radiation

The radiation balance of the Earth system is depicted in the Figure. (Shortwave radiation is colored purple and longwave radiation is in red.) Shortwave radiation from the Sun penetrates through space to the outer edge of the atmosphere unimpeded by the vacuum of outer space. If one places a surface oriented perpendicular to an incoming beam of light, 1.94 cal cm^{-2} min^{-1} of solar radiation will be received. This value is known as the *solar constant* but actually varies by a small amount as the Earth-Sun distance changes through the year. Once solar radiation begins to penetrate through the atmosphere this amount begins to decrease due to absorption and reflection.

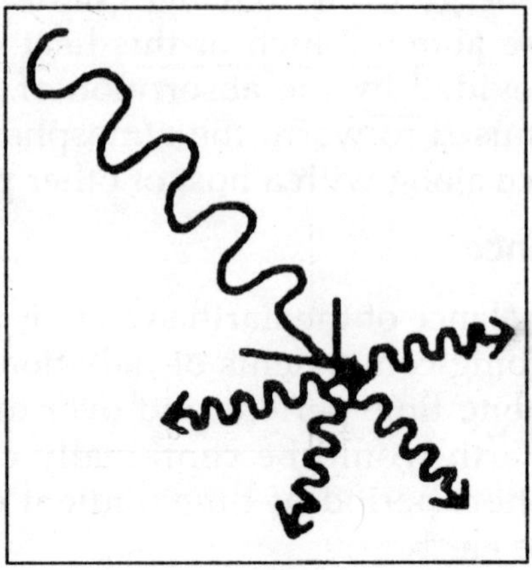

Fig. Scattering by Particles in the Atmosphere Causes a Beam of Light to be Broken into Several Weaker Beams of Light.

About 30% of the available solar radiation at the top of the atmosphere is reflected or scattered back to space by particulates and clouds before it reaches the ground. The gases of the atmosphere are relatively poor absorbers of solar radiation, absorbing only about 20% of what is available at the outer edge of the atmosphere. The remaining solar radiation makes its way to surface as direct and diffuse solar radiation.

Direct solar radiation (S) is shortwave radiation able to penetrate through the atmosphere without having been affected by constituents of the atmosphere in any way. *Diffuse radiation* (D) is shortwave radiation that has been scattered by gases in the atmosphere. Scattering is a process whereby a beam of radiation is broken down into many weaker rays redirected in other directions. Together, direct and diffuse shortwave radiation accounts for the total incoming solar radiation or *insolation* (K↓). In equation form:

$$K\downarrow = S+D$$

A portion of the incoming solar radiation is absorbed by the surface and a portion is also reflected away. The proportion of light reflected from a surface is the albedo (a). Albedo values range from 0 for no reflection to 1 for complete reflection of light striking the surface. Albedo can be expressed as a percentage (albedo multiplied by 100) that for some is easier to understand. For instance, grass has an albedo of about. This means that of the incoming solar radiation that strikes the grass, 23% of it is reflected away. On the other hand, highly reflective surfaces like snow have an albedo upwards of.87, or 87% of sunlight is reflected away.

Table ER. Sample Albedos

Surface	Typical Albedo
Fresh asphalt	0.04
Worn Asphalt	0.12
Bare Soil	0.17
Conifer forest (Summer)	0.08
Green Grass	0.25
New Concrete	0.55
Fresh Snow	0.80 - 0.90

The amount of reflection (K↑) is given by the following equation:

$$K\uparrow = (S+D)a$$

The amount of reflection basically depends on the colour of the surface, darker surfaces reflect less than lighter colored ones. For some surfaces, like water, the sun angle affects albedo. If we have been out on a lake during the day we might have noticed that the sun reflects off the surface more when the sun is lower in the sky than it does when it is more directly overhead.

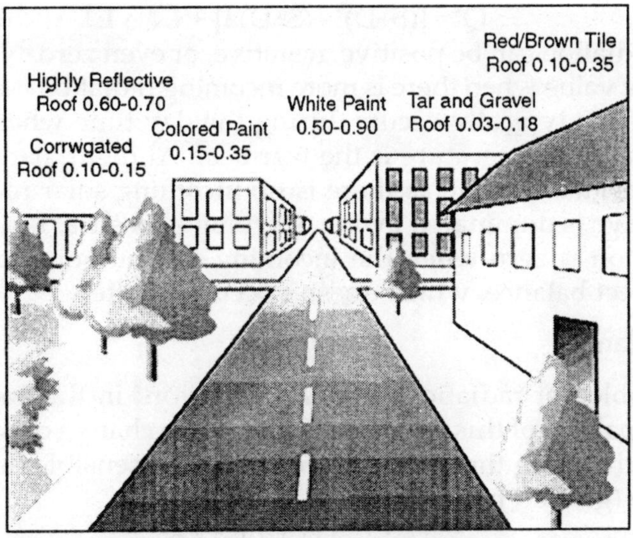

Fig. Darker Urban Surface Absorb More heat During the day Causing the Urban Heat Island Effect.

Net shortwave radiation is the difference between incoming and outgoing shortwave radiation expressed as:

$$K\uparrow (S+D) - (S+D)a$$

During the day, K* is a positive value as incoming always exceeds outgoing shortwave radiation. At night, K* is equal to zero as the Sun is below the horizon.

Longwave Radiation

The energy absorbed at the surface is radiated by the Earth as *terrestrial longwave radiation* (L↑). The amount of energy emitted is primarily dependent on the temperature of the surface. The hotter the surface the more radiant energy it will emit. The gases of the atmosphere are relatively good absorbers of longwave radiation and thus absorb the energy emitted by the Earth's surface. The absorbed radiation is emitted downward toward the surface as *longwave atmospheric counter-radiation* (L↓) keeping near surface temperatures warmer than they would be without this blanket of gases. This is known as the "greenhouse effect".

The difference between incoming and outgoing longwave radiation is net longwave radiation expressed as:

$$L^* = L\downarrow - L\uparrow$$

Net Radiation

Gathering all the radiation terms together we have net allwave radiation:

$$Q^* = [(S+D) - (S+D)a] + L\downarrow - L\uparrow$$

Net radiation can be positive, negative, or even zero. Net radiation is a positive value when there is more incoming radiation than outgoing radiation. This typically occurs during the day time when the sun is out and the air temperature is the warmest. At night, net radiation is usually a negative value as there is no incoming solar radiation and net longwave is dominated by the outgoing terrestrial longwave flux. Net radiation is zero when the incoming and outgoing components are in perfect balance, which doesn't occur too often.

Energy Balance

Available net radiation is used to do work in the Earth system. The principal use of this energy is in the phase change of water (latent heat, LE), changing the temperature of the air (sensible heat, H), and subsurface (ground heat, G) or,

$$Q^* = H + LE + G$$

Though there are other uses for net radiation like photosynthesis and the weathering of rocks, it is the three previously stated uses that are most important. LE, H, and G involve non-radiative transfers of heat. That is, conduction or convection/advection are responsible for the transfer of heat, not electromagnetic radiation.

When work is done, like heating the air above the surface, there is a transfer of energy (heat) from one place to the next. To illustrate the transfer of energy we'll use arrows either pointing away from or toward the surface of the Earth to indicate the direction of heat transfer. We will also use positive and negative signs to indicate that heat is being added to, or taken away from a body. Following the conventions of Sellers and Oke, non-radiative fluxes directed away from a surface are positive. Thus positive values indicates a loss of heat from the surface while negative values indicate a gain.

Sensible Heat Transfer(H)

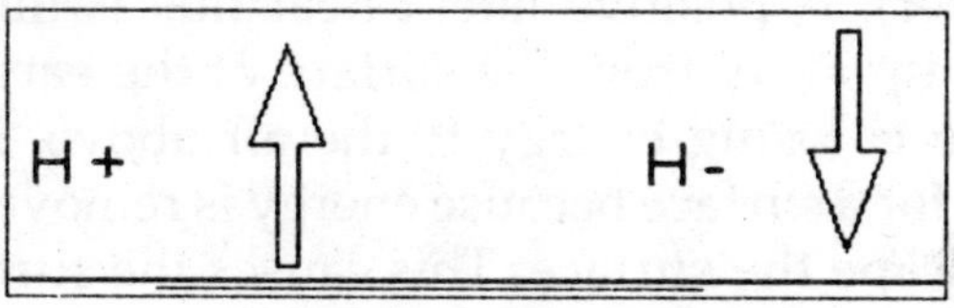

Fig. Sensible Heat Transfer

Sensible heat is heat energy transferred between the surface and air when there is a difference in temperature between them. A change in temperature over distance is called a "*temperature gradient*". In this case, it is a vertical temperature gradient, i.e., between the surface and the air above. We feel the transfer of sensible heat as a rise or fall in the temperature of the air. Heat is initially transferred into the air by conduction as air molecules collide with those of the surface. As the air warms it circulates upwards via convection. Thus the transfer of sensible heat is accomplished in a two-step process. Because air is such a poor conductor of heat, it is convection that is the most efficient way of transferring sensible heat into the air.

When the surface is warmer than the air above, heat will be transferred upwards into the air as a *positive sensible heat transfer*. The transfer of heat raises the air's temperature but cools the surface. If the air is warmer than the surface, heat is transferred from the air to the surface creating a *negative sensible heat transfer*. If heat is transferred out of the air, the air cools and the surface warms. This situation may take place at night when the sun goes down and there is no input of solar radiation. At this time, the ground cools due to longwave emission and the air directly above the surface is warmer.

Latent Heat Transfer (LE)

When energy is added to water it will change states or phase. The phase change of a liquid to a gas is called *evaporation*. If we could see down to the molecular level we would find water being comprised of cluster of water molecules (H_2O). The clusters are bound together by bonding between the hydrogen atoms of water molecules. The heat added during evaporation breaks the bonds between the clusters creating individual molecules that escape the surface as a gas. The heat used in the phase change from a liquid to a gas is called the *latent heat of vaporization*. We say it is "latent" because it is being stored in the water molecules to later be released during the condensation process. We can't sense or feel latent heat as it does not raise the temperature of the water molecules.

When evaporation is taking place we say there is a *positive latent heat flux* (transfer). A positive latent heat flux is illustrated with an arrow pointing up away from the surface of the earth. This indicates that the surface is losing energy to the air above. Evaporation is a cooling process for a surface because energy is removed from the water as molecules escape the surface. This causes the surface temperature to decrease. We 've probably experienced this cooling when water or sweat evaporates from our skin.

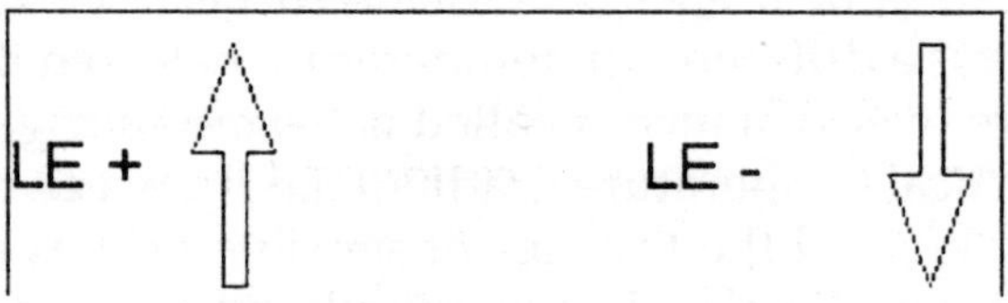

Fig. Latent Energy Transfer

Condensation is the phase change from a gas to a liquid. During the phase change, the latent heat that was taken up during evaporation is released from the water molecule and passed into the surrounding air. During this process latent heat is converted to sensible heat causing an increase in the temperature of the air.

When radiation is absorbed by the Earth it will raise the temperature of the surface. But if the surface is water, some of that energy is used in evaporation rather than heating the water. As a result, with equal inputs of energy to land and water surfaces, land will heat up more than water. This is one of the reasons why it is cooler near large bodies of water.

Ground Heat Transfer (G)

The third major use of radiant energy is to warm the subsurface of the Earth. Heat is transferred from the surface downwards via conduction. Like in the case of sensible heat transfer, a temperature gradient must exist between the surface and the subsurface for heat transfer to occur. Heat is transferred downwards when the surface is warmer than the subsurface (positive ground heat flux). If the subsurface is warmer than the surface then heat is transferred upwards (negative ground heat flux). The transfer of energy to and from the surface varies over the course of a day. Figure shows the relative amounts and direction of energy flow during the day and night for a moist surface. During the day net radiation is a positive value as incoming radiation exceeds outgoing radiation allowing the surface to gain energy. The energy is distributed over the three major categories of energy use, LE, H, and G. During the day, the available radiant energy is used to evaporate water into the air, raising the air's humidity. Sensible heat is transferred upwards to warm the air above the surface.

Heat is also conducted down into the subsurface. At night the processes reverse. At night with no incoming solar radiation there is more outgoing radiation than incoming creating a negative value for net radiation. Under these circumstances the surface cools due to a loss of energy and heat is transferred from the air toward the surface. As air cools through the evening the loss of energy allow condensation to occur, so long as the air's humidity is at or near saturation. Notice that the arrow for the ground heat flux G is pointing upward toward the surface. This indicates that the energy that had been stored in the subsurface during the day is now conducted toward the surface. This occurs as the surface is cooler than the subsurface.

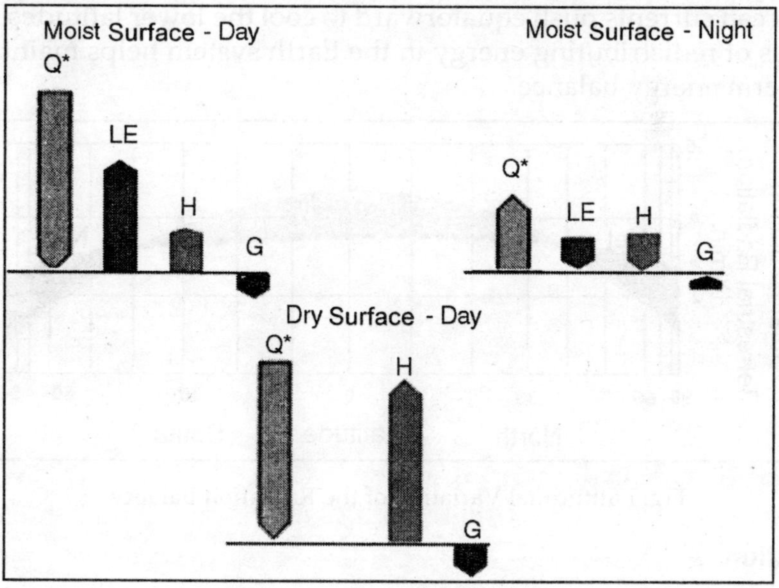

Fig. Types of Energy Balances.

Figure also shows the differences between how energy is used in moist and dry locations. Water is available at the surface for evaporation and latent heat transfer into the air at moist locations. Without available water, no transfer of latent energy occurs, hence the absence of an LE flux for the dry surface. Most of the available energy, Q*, is allocated to sensible heat transfer creating warm air temperatures.

GLOBAL PATTERNS OF INSOLATION

Patterns of insolation and net radiation which determine the location of plants, animals, climate, soils and other elements of our physical environment can be discerned from Figures. Figure illustrates the latitudinal distribution of incoming solar radiation and outgoing terrestrial radiation. From approximately 35° N to 35° S latitude there is a surplus of energy as incoming radiation exceeds outgoing. The blue regions indicate that there is more outgoing energy than incoming, yielding a net loss of energy from the Earth's surface.

One might ask then why the middle to higher latitudes aren't getting colder through time as a result of the net loss, and the subtropical to equatorial regions getting constantly hotter due to the net gain. The reason is that the energy is redistributed by circulation of the atmosphere and oceans. Heat gained in the tropics is transported poleward by the global circulation of air and warm ocean currents to heat higher latitude regions. Cooler air from the higher latitudes and cold ocean currents push equatorward to cool the lower latitudes. This process of redistributing energy in the Earth system helps maintain a long-term energy balance.

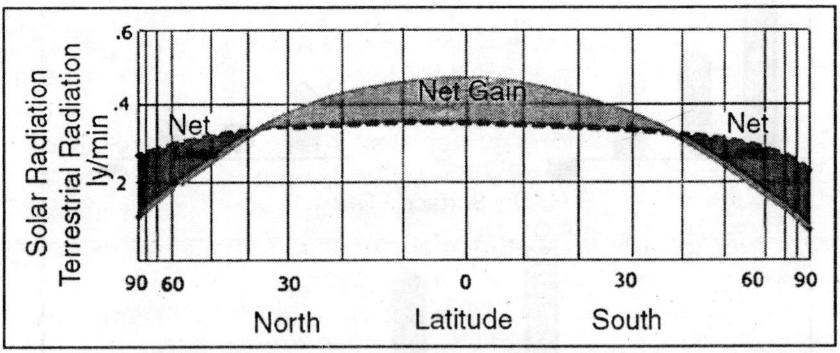

Fig. Latitudinal Variation of the Radiation Balance

Insolation

Figure simplifies the geographic distribution of insolation. For the Earth as a whole, particular patterns can be accounted for by variation in surface features that impact insolation. Insolation maxima are found in the tropical and subtropical deserts of the earth. Here, high sun angles and a lack of cloud cover in desert regions allow much solar radiation to the surface. Insolation decreases to a minimum at the poles where low sun angles and the Sun doesn't rise above the horizon nearly half the year reduces annual insolation.

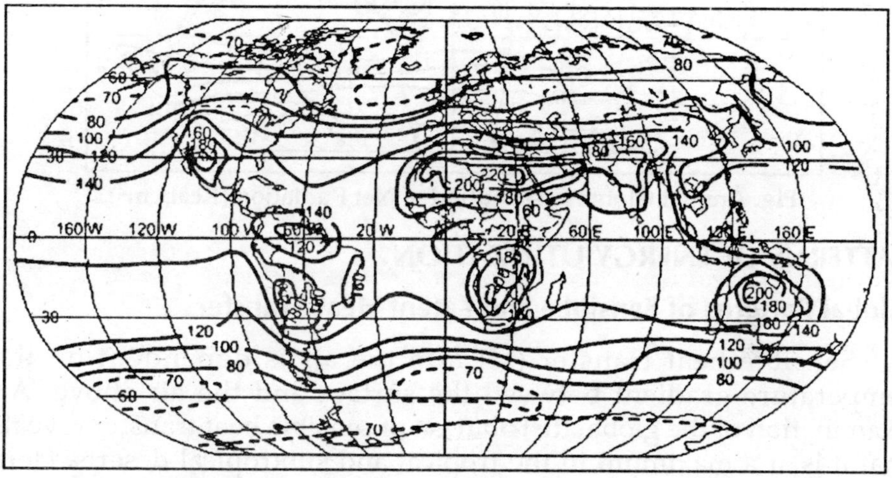

Fig. Annual Global Distribution of Solar Radiation (Kcal/cm^2)

Net Radiation

Net radiation exhibits a different pattern from that of insolation. Maximum net radiation is found over the tropical and subtropical oceans. The sun angle is always high over the tropical oceans so the surface receives intense radiation throughout the year. With an high sun angle the albedo of the surface is low and absorption is high. However, the energy received is partitioned into warming the surface as well as evaporating water. The result is a lower surface temperature than one might expect with such high sun angles.

Additionally, the high specific heat of water means that it takes much more energy to heat a unit mass of water than that of land, resulting in lower ocean surface temperatures. With lower surface temperatures the water surface does not radiate as much longwave radiation out to the atmosphere as nearby land at the same latitude. With much radiation coming in and little going out, the net value is large compared to land at the same latitude. Net radiation is at a minimum over the poles as the sunlight that comes in at a low angle is reflected from the ice-covered surface. Combined with the long polar night, very little net radiation is found at these latitudes.

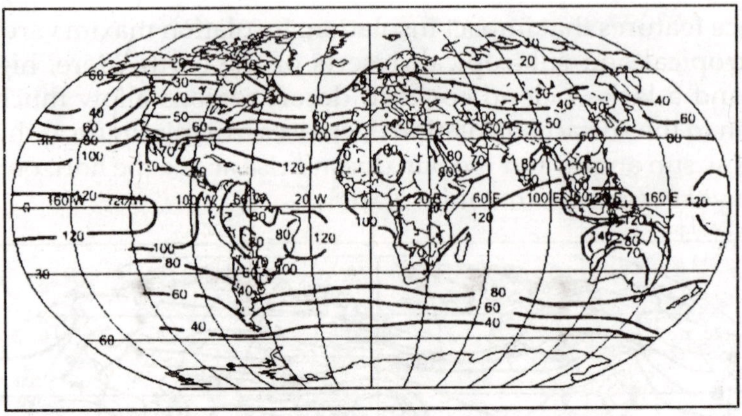

Fig. Annual Global Distribution of Net Radiation (Kcal/cm^2)

PATTERNS OF ENERGY UTILIZATION

Global Patterns of SensibLe and Latent Heat Transfer

Sensible heat transfer (H) into the air is dependent on the temperature gradient between the surface and the air above. An examination of the global distribution of sensible heat transfer reveals that it is at a maximum in the tropical and subtropical deserts. Here the high surface temperature conducts much heat into the air above. Sensible heat is lowest near the poles where the surface temperatures are quite cold.

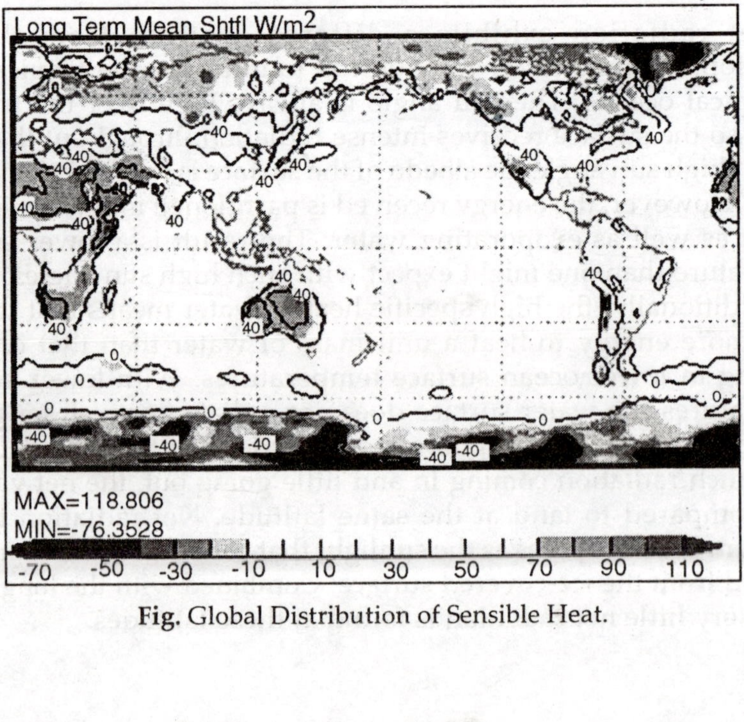

Fig. Global Distribution of Sensible Heat.

The maximum amount of latent heat transfer (LE) occurs over the subtropical oceans where there is a maximum of net radiation and, of course, water to evaporate. The lowest rates occur in desert locations. Even though there is ample available energy for LE, there is little water present.

Off the east coast of mid latitude continents like North America, we'll find very high rates of latent energy transfer into the air. Here, dry air blows off the continent and over the warm ocean current that flows along the coast. This produces a large moisture gradient between the surface and the air above that induces evaporation and LE transfer.

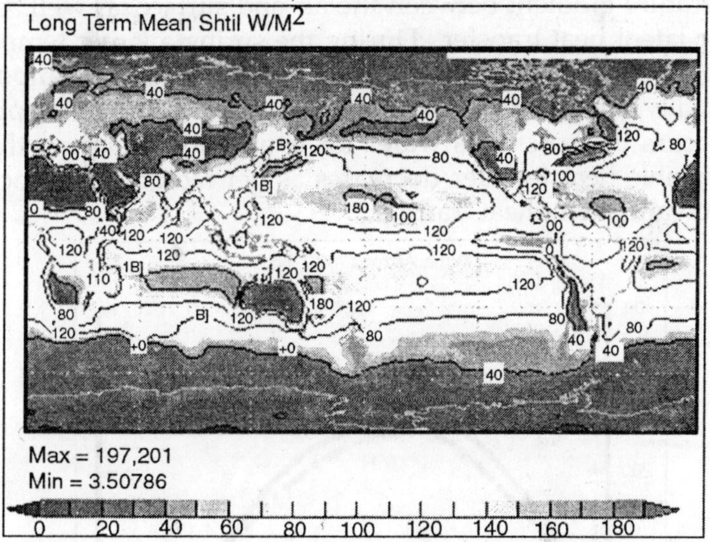

Fig. Global Distribution of Latent Heat

Regional Patterns of Sensible and Latent Heat Transfer

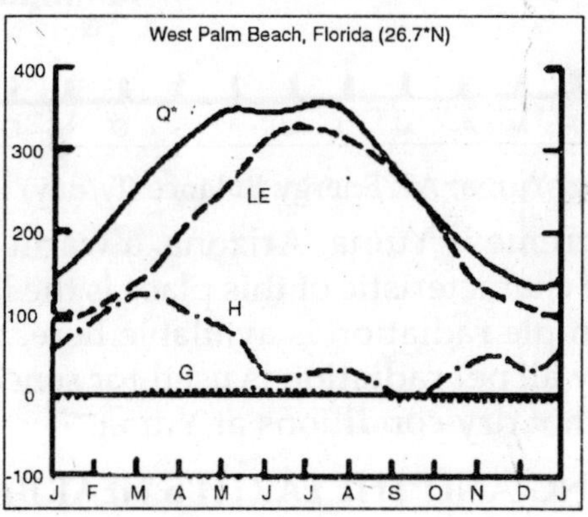

Fig. West Palm Beach, Fl Energy Balance (ly/day)

West Palm Beach, Florida is located in a warm and moist climate.

The annual variation of net radiation shows a typical pattern of Q* being at a maximum during the summer and minimum during the winter. Latent heat transfer (LE) is high at West Palm Beach because of the water availability near the ocean. Latent energy transfer into the air is greatest during the summer time which is the wettest period of the year, and when net radiation is the highest.

Warmer summer time air can hold more moisture and hence there is a larger moisture gradient to drive latent heat transfer. The annual variation of sensible heat transfer is determined by the available net radiation and the temperature gradient between the air and surface, as well as use of energy for latent heat transfer. During the spring a larger temperature gradient between the surface and the air above exists. More important is the availability of water for evaporation. During the summer, appreciable rainfall occurs that provides water for evaporation. During this time period, sensible heat transfer decreases as net radiation is allocated to evaporation and latent heat transfer.

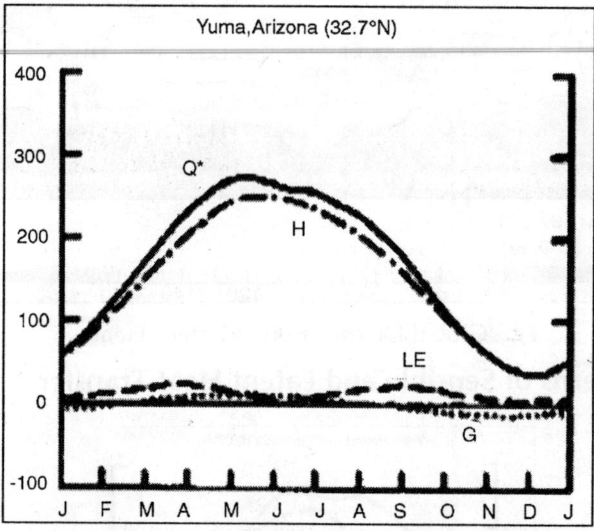

Fig. Yuma, AZ Energy Balance (ly/day)

At the other extreme is Yuma, Arizona, a warm and dry climate. The most noticeable characteristic of this place is the lack of latent heat transfer. Though ample radiation is available here, there is no water to evaporate. Nearly all net radiation is used for sensible heat transfer which explains the hot dry conditions at Yuma.

RADIATIVE FORCING AND THE EARTH'S HEAT BALANCE

The future physical geography of Earth as affected by global warming comes down to changes in the heat balance of the Earth system. Radiative forcing is a measure of the strength of agents, both natural and human, that cause climate change. Radiative forcing agents are factors that change the balance between incoming solar radiation and outgoing infrared radiation within the Earth's atmosphere. The radiative forcing of greenhouse gases is what will propel much of this change. Documented changes with relatively high level of scientific understanding from 1750 to 2005 has been presented by the Intergovernmental Panel on Climate Change.

The Present Picture

Two long-lived greenhouse gases produced from human activities, carbon dioxide and methane, are the most potent, contributing 1.66 and 0.48 (W/m^2) respectively at the global scale. Shorter-lived tropospheric ozone contributes an average of 0.35 (W/m^2), though we have only a medium level of scientific understanding of its contribution.

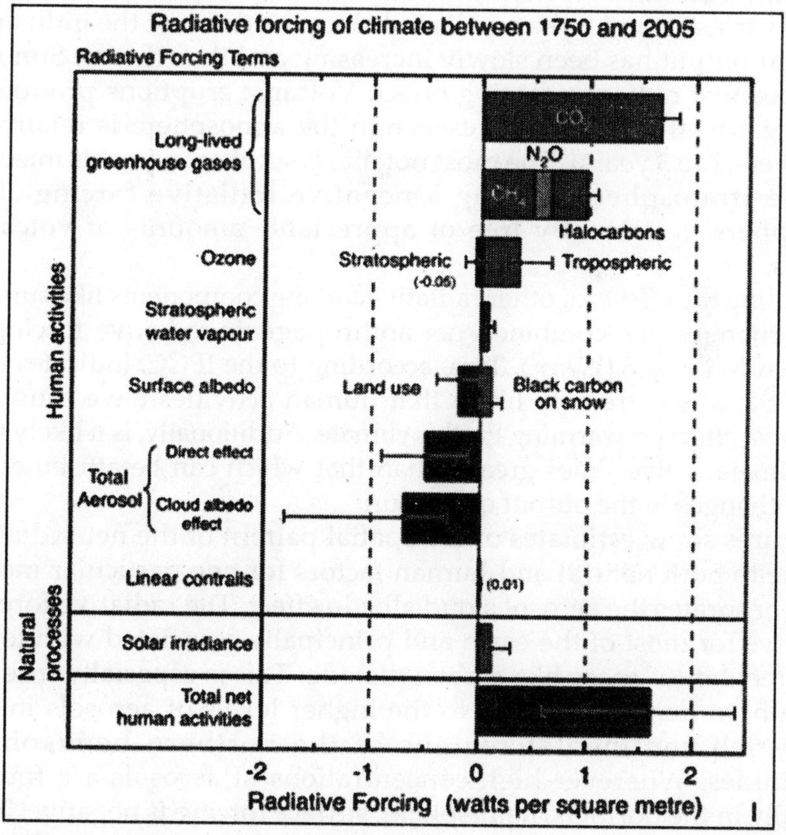

Fig. Principal Agents of the Radiative Forcing of Climate Change 1750 - 2005.

Though human activities affect agents that increase radiative forcing, they also negatively impact radiative forcing. A notable impact, is that of the aerosol content of the atmosphere. Aerosols themselves can negatively impact radiative forcing by reflecting solar radiation. But a much larger impact, and one with less scientific certainty, is their role in creating clouds which also reflect solar radiation.

In addition to changes in the gaseous composition of the atmosphere, surface albedo changes from human activities also effects radiative forcing. This is particularly true when forests are cleared for agriculture. Forests generally have a lower albedo than open land thus absorbing more incident solar radiation. These changes induce a radiative forcing by effecting the shortwave radiation balance. This is particularly true when snow is present. Open land has a more complete cover of highly reflective snow, while the lower albedo trees stand above the snow.

Natural radiative forcing largely results from changes in solar output and volcanic eruptions. As noted earlier in this chapter, sunspot activity varies on an 11-year cycle. But since the dawn of the industrial age, solar output has been slowly increasing and thus the the Sun has had a positive radiative forcing effect. Volcanic eruptions produce a variety gases, though their presence in the atmosphere is relatively short lived (2 to 3 years). The most notable is sulphate aerosols injected into the stratosphere causing a negative radiative forcing. The stratosphere is presently free of appreciable amounts of volcanic aerosols.

Adding the effects of other radiative forcing components like surface albedo changes, the combined net anthropogenic radiative forcing is estimated to be +1.6 (W/m^2). This, according to the IPCC, indicates that since 1750, it is extremely likely that human activities have caused a substantial effect on warming Earth's climate. Additionally, is it likely that this estimate is five times greater than that which can be attributed to natural changes in the output of the Sun.

Figures show estimates of the spatial pattern of the net radiative flux due to both natural and human factors for one particular model that incorporates the aerosol cloud albedo effect. The radiative forcing is positive for most of the earth and principally associated with long-lived greenhouse gases like carbon dioxide. This is especially true for the southern hemisphere due to the higher levels of aerosols in the source-rich continental regions of the northern hemisphere midlatitudes. Wherever high concentrations of aerosols are found, especially in the northern hemisphere, surface forcing is negative. This is due to the influence of aerosols in reducing the shortwave radiation that reaches the ground. As aerosol concentrations are lower over much of the southern hemisphere oceans and at high latitudes. With less aerosols more shortwave radiation reaches the earth thus surface forcing becomes positive.

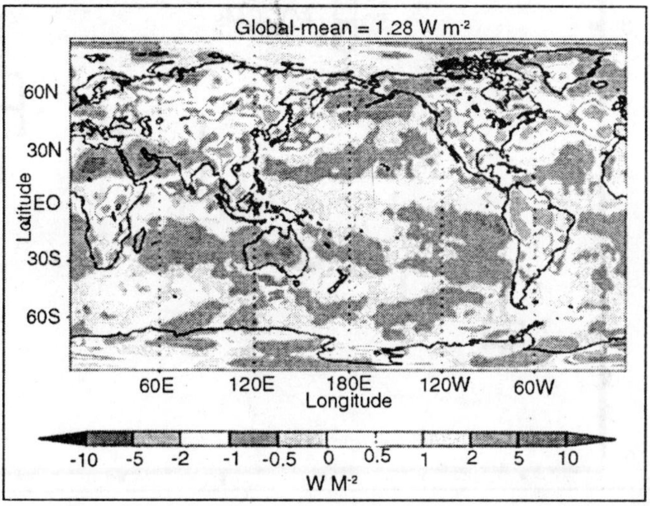

Fig. Instantaneous Change in the Distribution of the Net Radiative Flux (W/m^2) due to Natural Plus Anthropogenic Radiative Forcings Between the Years 1860 and 2000 at the Tropopause.

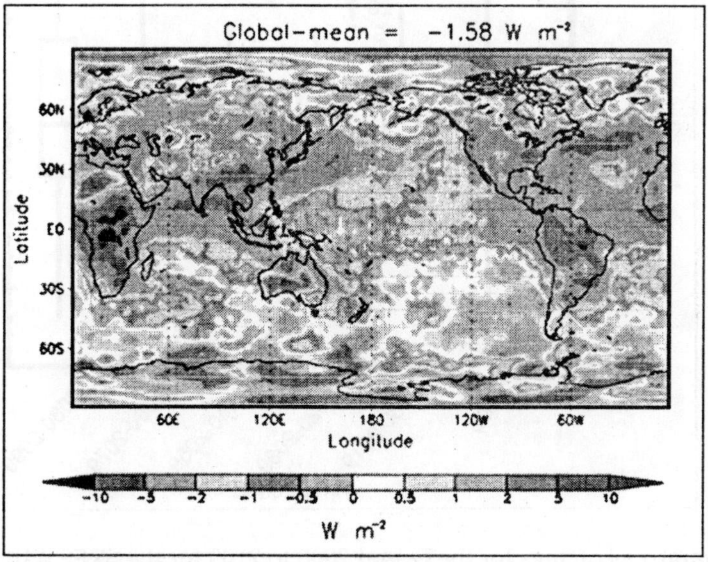

Fig. Instantaneous Change in the Distribution of the net Radiative Flux (W/m^2) due to Natural Plus Anthropogenic Surface Forcings Between the Years 1860 and 2000.

Future Forcings

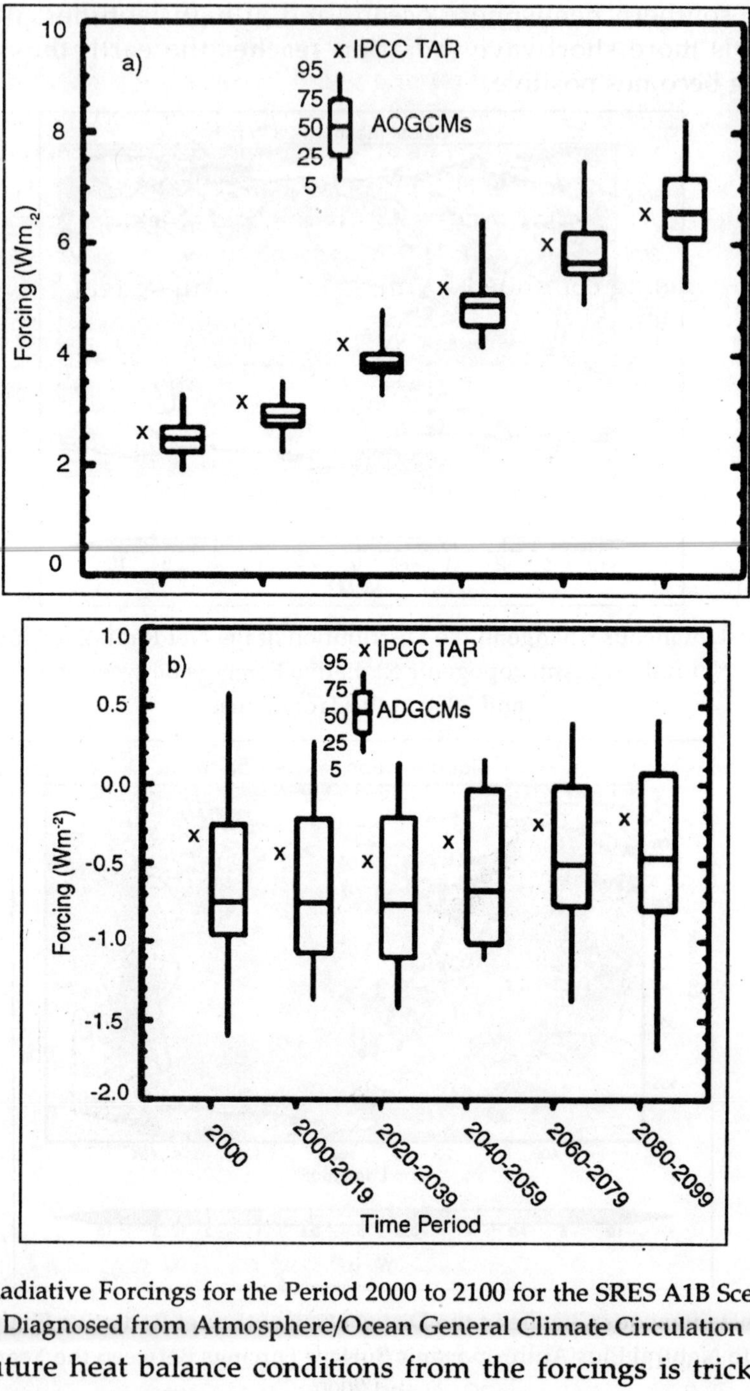

Fig. Radiative Forcings for the Period 2000 to 2100 for the SRES A1B Scenario Diagnosed from Atmosphere/Ocean General Climate Circulation

Future heat balance conditions from the forcings is tricky and predictions from various AOGCMs do not precisely agree with one another, largely due to the radiatively active gases used, solar variability, land use change, and how radiation transfer is formulated. Figure shows the range of forcing using "box and whisker" diagrams estimated by these models using a scenario where economic growth is rapid, world population peaks at 9 billion in 2050 and then declines, there is a quick and efficient spread of technologies, extensive social and cultural interaction occurs worldwide, and world-income and way of life converges between regions. Longwave forcing continues to increase fueling continued warming of the earth system.

Global warming will continue unabated without changes in how human activities impact radiative forcing agents. In future chapters, we'll investigate how changes in these agents will impact patterns of temperature, pressure and wind, weather, biotic systems, the hydrosphere and earth surface processes.

Chapter 4

Geomagnetism

The Earth's magnetic field is both expansive and complicated. It is generated by electric currents that are deep within the Earth and high above the surface. All of these currents contribute to the total geomagnetic field. In some ways, one can consider the Earth's magnetic field, measured at a particular instance and at a particular location, to be the superposition of symptoms of a myriad of physical processes occurring everywhere else in the world. The challenge is to untangle the rich information content of the magnetic field so that we can better understand our planet and the surrounding space environment in which it resides.

Obviously, it would be a daunting task to try to summarize every single aspect of the subject of geomagnetism. Indeed, the necessity for brevity here means that our exposition will omit some relatively enormous details.

MAGNETIC-FIELD

Perhaps the most familiar demonstration of the reality of the Earth's magnetic field is the north-seeking tendency of the compass needle, a property that has been exploited by navigators for centuries. A compass is constructed, of course, such that its magnetic needle is free to rotate in the horizontal plane. But if we were to permit the compass needle to have full directional freedom, suspending the needle from a thread so that it could freely obtain its orientation both horizontally and vertically, we would find that the alignment of the needle would vary continuously from one point in space to another.

Furthermore, if we were to measure the force on the magnetic needle causing it to assume its preferred alignment, we would find that the strength of this force, proportional to the intensity of the magnetic field, also varies continuously with position in space. These properties can be used to map a continuous family of lines of force — vectors having both direction and magnitude. At any one individual point on the surface of the Earth the orientation of the geomagnetic vector is conventionally described relative to a geographic coordinate system using two angles: declination, the angle between the horizontal component of the magnetic-field vector relative to true north, and inclination, the angle between the horizontal plane and the total field vector. The intensity of the magnetic field, which is independent of the orientation of the reference coordinate system, is represented by the length of the vector.

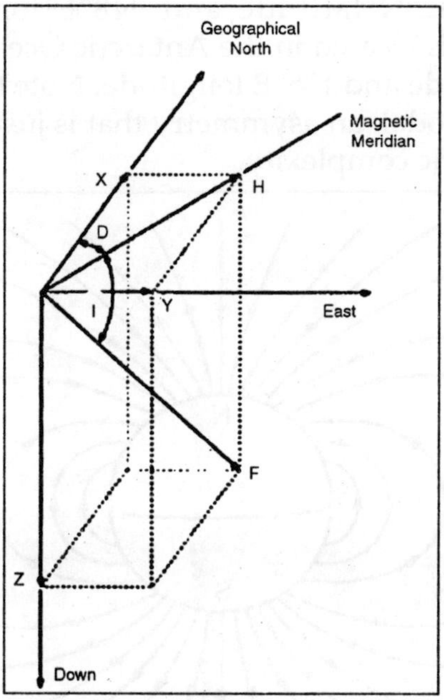

Fig. Magnetic Field Components

THE SPATIAL FORM OF THE GEOMAGNETIC FIELD

The magnetic field of the Earth is often times described as being approximately dipolar, with field lines emanating from the south geomagnetic pole and converging at the north geomagnetic pole, as depicted in the figure below. Although this description is useful for many purposes, it is not particularly accurate. The dipolar part of the field is actually tilted by approximately 11° with respect to the rotational axis, and there are additional, non-dipolar ingredients in the geomagnetic field, all of which, when added together, are the total surficial field in all of its complex detail. As a result of this complexity, not only does the direction of the compass needle deviate from true north, but the amount of the deviation, the declination, varies as a function of geographic location.

This fact has been an historical nuisance for navigators, and, not surprisingly, it helped to motive some of the original global-scale surveys of the Earth's magnetic field. Another simple measure of the field's geometry is the position of the magnetic poles. At the north geomagnetic pole, our freely moving magnetic needle would point down, whilst at the south geomagnetic pole, the needle would point up. For these reasons, the geomagnetic poles are sometimes referred to as 'dip poles'. The north geomagnetic pole is located in the Canadian Arctic at about 82°N latitude and 248°E longitude. The south geomagnetic pole is located in the Antarctic Ocean south of Australia at about 65°S latitude and 138°E longitude. Note that the geomagnetic poles are not antipodal, an asymmetry that is just another measure of the field's geometric complexity.

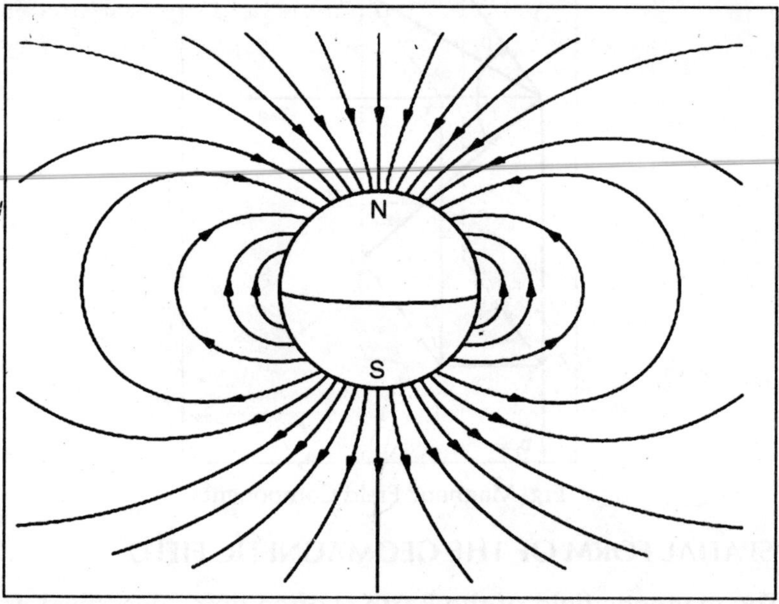

Fig. Dipole field

THE GEODYNAMO

The Earth is, of course, extremely complicated; it consists of many different interacting parts. But broadly speaking the Earth below our feet is stratified in radius, being composed of a solid-iron inner core, a liquid-iron outer core, and an electrically-insulating, rocky over-lying mantle. The main part of the Earth's magnetic field is generated by electric currents sustained by a dynamo situated in the core, and the study of the form and long-term behaviour of the geomagnetic field can be used to discover how the geodynamo works. Paleomagnetic measurements of rocks indicate that the Earth has possessed a magnetic field for at least 3.5 billion years, and yet, without some sort of regenerative process to offset the inevitable ohmic dissipation of electric currents, the geomagnetic field would vanish in about 15,000 years.

Therefore, the dynamo in the core must be regenerative, and it is generally thought that this regenerative process relies on the principles of magnetic induction. In effect, the core is a naturally occurring electric generator, where convection kinetic energy, driven by chemical differentiation and the heat of internal radioactivity, is converted into electrical-magnetic energy. More specifically, electrically-conducting fluid flowing across magnetic-field lines induces an electric current, and this generated current supports its own associated magnetic field. Depending on the geometrical relationship between the fluid flow and the magnetic field, the generated magnetic field can reinforce the pre-existing magentic field, in which case the dynamo is said to be 'self-sustaining'.

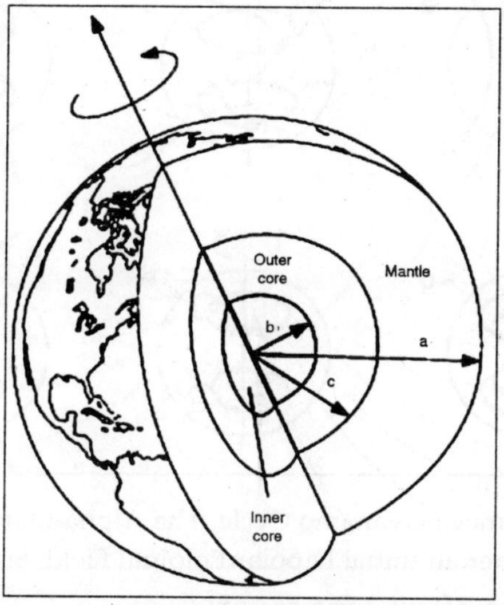

Fig. The Interior of the Earth. The anatomy of the Earth. The Mantle has a Radius

- Of 6371 km, the inner core has a radius
- Of 1215 km, and the outer core has a radius
- Of 3485 km.

The details of exactly how the dynamo works are not entirely resolved. Hence, the theory is the subject of on-going research. Nonetheless, a reasonably coherent qualitative understanding exists of how the core's fluid motion sustains the Earth's magnetic field. We know that the dynamo is governed by dynamical, nonlinear mathematical equations, somewhat akin to the equations of meteorology and oceanography, but with the additional complication presented by the magnetic field itself.

From a purely kinematic standpoint dynamo action relies on the so-called 'alpha-omega' process, in which core fluid motion, influenced by the Coriolis force, consists of a combination of differential rotation and convective, turbulent helical motion. These two motional ingredients work together to reinforce the magnetic field and, thereby, offset the destructive affects of ohmic dissipation. The alpha-omega process successfully describes how it is that the magnetic field can be amplified, but it is the dynamics that ultimately determine the field's strength: the field grows until a rough balance is attained between the Coriolis and the Lorentz forces.

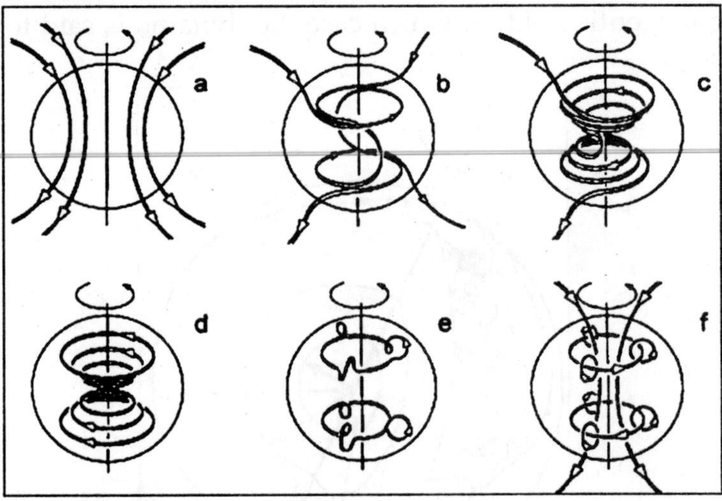

Fig. The Alpha-Omega Dyanamo Cycle. The Alpha-Omega Dynamo Cycle: Consider an Initial Dipolar Poloidal Field, Such as in

- The omega-effect consists of
- Differential rotation, wrapping the magnetic field around the rotational axis, thereby creating
- A quadrupolar toroidal field magnetic field inside the core.
- A closure of the dynamo cycle requires a bit of symmetry breaking, brought about by the alpha-effect, whereby
- Helical upwelling and downwelling creates loops of magnetic field. These loops coalesce
- To reinforce the original dipolar field.

FIELD VARIATION OVER HISTORICAL TIMESCALES

Although the magnetic field is sustained by the dynamo within the Earth's core, part of the field threads its way through the mantle and up to the surface, where it has influence on compasses and where it can be measured by magnetometers. The same convective motion that drives the geodynamo also causes the field, measured at the surface, to be time-dependent over historical timescales, a phenomenon known as 'secular variation'. In fact, magnetic models and charts must be periodically updated to accommodate the continual secular variation of the field. Secular variation can also be clearly seen in long-timescale, magnetic records collected at ground-based observatories.

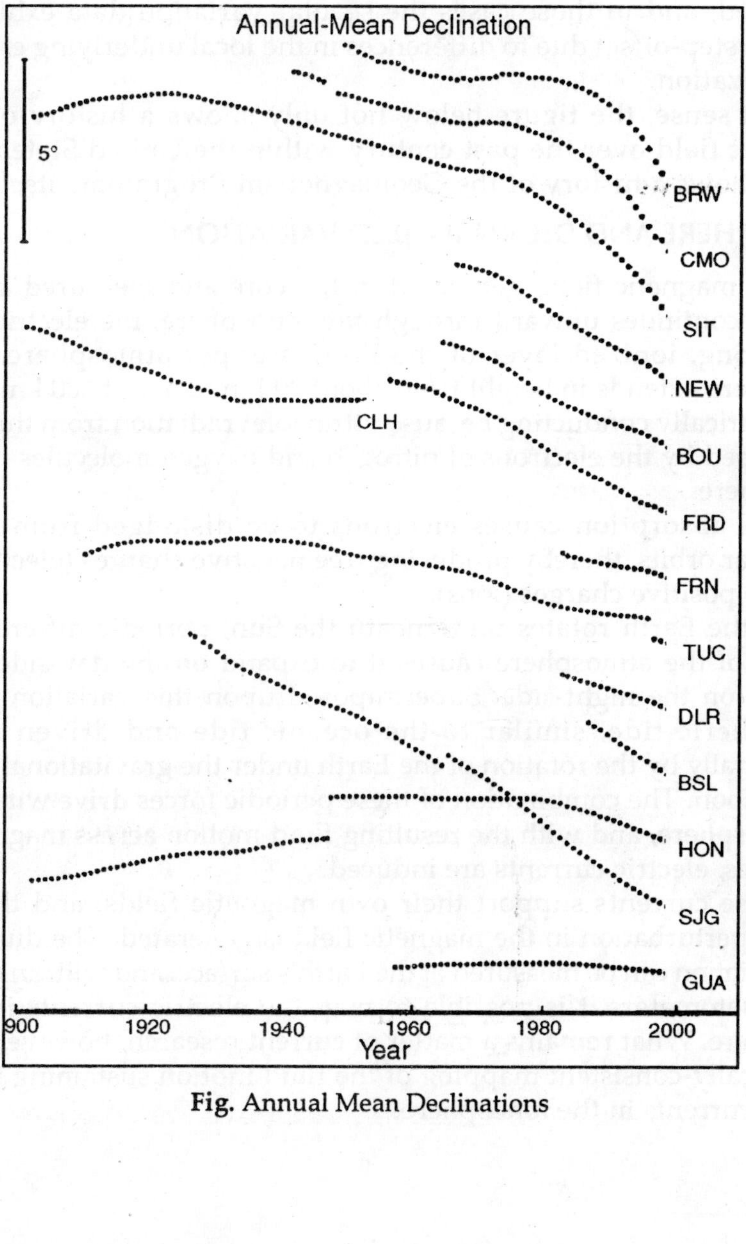

Fig. Annual Mean Declinations

In the figure shows annual-mean declinations take from over a century's worth of data collected from the various USGS Geomagnetism Programme observatories. For making long-term studies of the secular variation it is important that magnetic measurements such as these be made relative to a stable reference. In the case of the observatories, this means that the same stable absolutes pier must be used year-after-year.

Of course, for practical reasons, occasionally, an observatory pier must be moved, and sometimes even an entire observatory must be moved, and in those cases the secular variation data exhibit a baseline step-offset due to differences in the local underlying crustal magnetization.

In a sense, the figure below not only shows a history of the magnetic field over the past century within the United States, but it also shows a history of the Geomagnetism Programme itself.

IONOSPHERE AND DIURNAL-FIELD VARIATION

The magnetic field, generated in the core and measured at the surface, continues upward through the ionosphere, the electrically-conducting, ionized layer of the Earth's upper atmosphere. The ionosphere extends in height from about 90 km to about 600 km, and it is electrically conducting because ultraviolet radiation from the Sun is absorbed by the electrons of nitrogen and oxygen molecules in the atmosphere.

This absorption causes electrons to be dislodged from their molecular orbits, thereby producing free negative charges (electrons) and free positive charges (ions).

As the Earth rotates underneath the Sun, periodic differential heating of the atmosphere causes it to expand on the day-side and contract on the night-side. Superimposed upon this variation is an atmospheric tide, similar to the oceanic tide and driven most substantially by the rotation of the Earth under the gravitational field of the Moon. The combination of these periodic forces drive winds in the ionosphere, and with the resulting fluid motion across magnetic-field lines, electric currents are induced.

These currents support their own magnetic fields, and thus a diurnal perturbation in the magnetic field is generated. The diurnal-field variation can be measured at the Earth's surface, and with an array of magnetometers it is possible to map the electric currents in the ionosphere. What remains a matter of current research, however, is a dynamically-consistent mapping of the fluid motion sustaining these electric currents in the ionosphere.

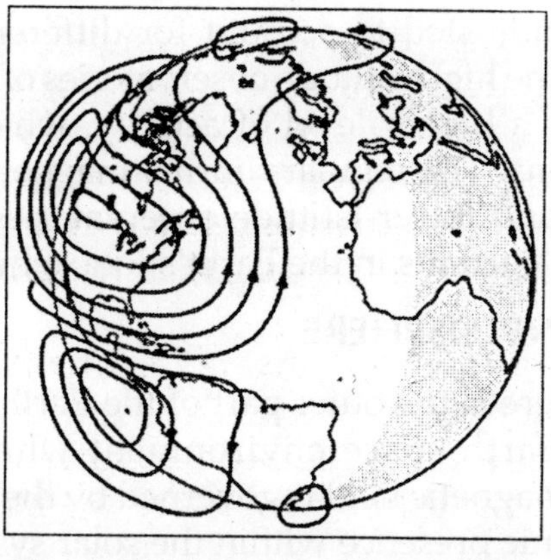

Fig. Ionospheric Electric Currents

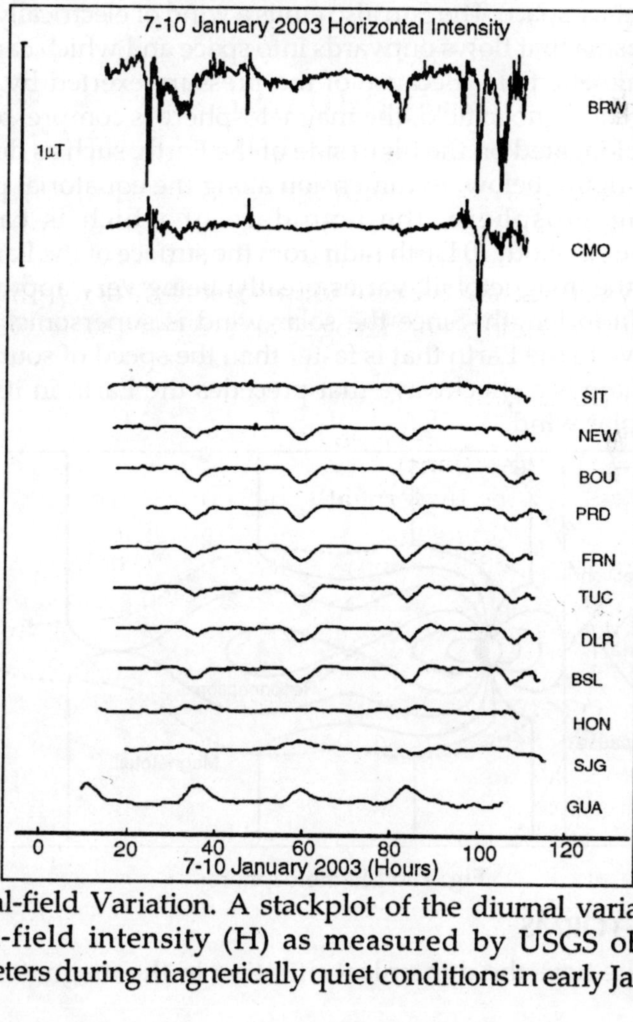

Diurnal-field Variation. A stackplot of the diurnal variation in the horizontal-field intensity (H) as measured by USGS observatory magnetometers during magnetically quiet conditions in early January 2003.

The data have been adjusted to account for different observatory site longitudes. Note that the high-latitude observatories of Barrow (BRW) and College (CMO) show a higher level of activity, which is normal, even during quiet conditions. The regular diurnal magnetic-field variation is seen clearly as mid- and lower-latitude observatories, and is generated by large-scale electric currents in the Earth's ionosphere.

THE EARTH'S MAGNETOSPHERE

The magnetosphere is the outer part of the Earth's magnetic field, a region in the near-Earth space environment where the shape and behaviour of the geomagnetic field is governed by the Sun. Of course, the Sun is a highly dynamic presence within the solar system. It has its own dynamo, generating a somewhat tangled magnetic field that extends out into interplanetary space. The Sun also emits a wind of electrically-charged particles, a plasma that flows outwards into space and which carries with it the heliomagnetic field. Because of the pressure exerted by the solar wind on the geomagnetic field, the magnetosphere is compressed on the day side and elongated on the night side of the Earth, such as depicted in the schematic figure below. In dimension along the equatorial plane, the day-side magnetosphere, the boundary of which is called the 'magnetopause', is about 10 Earth radii from the surface of the Earth, while the length of the 'magnetotail' varies greatly, being very approximately 100 Earth radii in length. Since the solar wind is supersonic, having a velocity relative to the Earth that is faster than the speed of sound within the plasma, there is a shockwave that precedes the Earth in its passage through the solar wind.

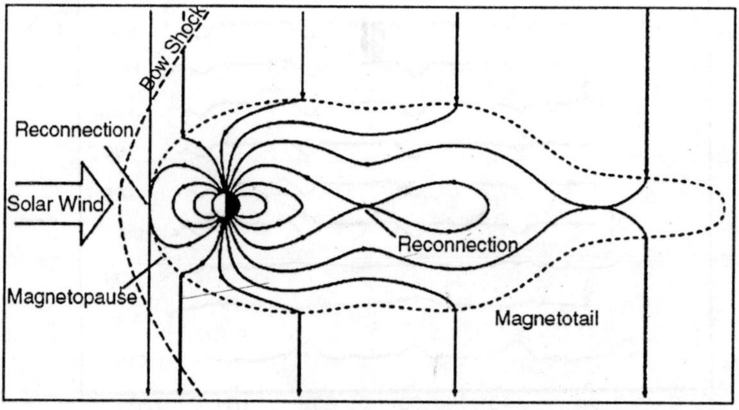

Fig. The Magnetosphere

MAGNETIC STORMS

Most of the time the dynamical situation in the magnetosphere is quasi-stationary, with little effect on (say) ground-based magnetometer measurements. Occasionally, however, the magnetosphere is extremely time-dependent, having an inductive enhancement of magnetospheric electric currents and their associated magnetic fields. This is manifest in ground-based magnetometer measurements as rapid, erratic variation. In analogy with atmospheric meteorology, such periods of magnetic activity are known as 'magnetic storms', and they are triggered by the interplanetary conditions established by the Sun.

The pressure balance between the solar wind and the geomagnetic field is delicate, and perturbations in solar-wind velocity can cause the magnetosphere to oscillate. Even more dramatically, occasionally the Sun emits a sudden gust of solar wind, a so-called 'coronal mass ejection'. If this impacts upon the magnetosphere then a magnetic storm can follow. Alternatively, magnetic storms can also be caused by a process called 'magnetic reconnection'. This is important in two places, in the magnetopause and in the magnetotail, and it is illustrated in the figure above. Let's first consider the reconnective process at the magnetopause. In the complicated flow of the solar wind, the interplanetary magnetic field can take on virtually any orientation, but sometimes it assumes a southward orientation.

In this orientation the interplanetary field has the opposite orientation of the Earth's magnetic field at the magnetopause. This juxtaposition of oppositely-oriented field lines is an unstable situation, one which mathematically is identical to a shear-flow instability in fluid mechanics. With a little bit of diffusion, brought about by electrical resistance, the interplanetary magnetic field can connect onto the Earth's field. Then, particles from the Sun, which would otherwise be deflected by the Earth's field, can stream along these continuously-connected field lines and enter into the magnetospheric cavity.

Further advection of the interplanetary field lines by the solar wind peels back the geomagnetic field, and the tail of the magnetosphere is further stretched as well. In the equatorial plane of the magnetotail, and with this stretching of the field lines, we have a second place where neighboring field lines have opposite orientation, see, once again, the figure above.

As before, with a little diffusion we obtain reconnection, causing part of the magnetic field, and the plasma in which it is embedded, to break off from the rest of the geomagnetic field and float down the tail. It is rather like the boudinage familiar from extensional geological formations! Simultaneously, the inner field lines recoil and accelerate the plasma particles within the magnetosphere, giving rise to an enhancement of electric currents. All of this, of course, describes a highly dynamic process, which causes the magnetic field measured at the Earth's surface to become extremely active.

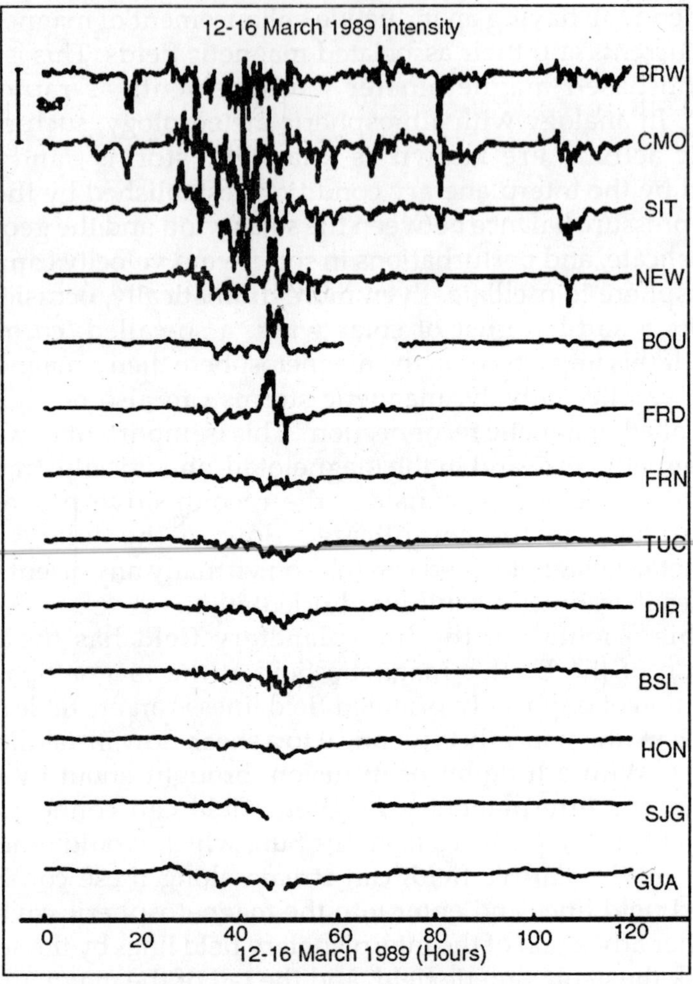

Fig. March 1989 Storm Data

EARTH'S MAGNETIC FIELD

Earth's magnetic field (and the surface magnetic field) is approximately a magnetic dipole, with one pole near the north pole and the other near the geographic south pole. An imaginary line joining the magnetic poles would be inclined by approximately 11.3° from the planet's axis of rotation. The cause of the field can be explained by dynamo theory.

Magnetic fields extend infinitely, though they are weaker further from their source. The Earth's magnetic field, which effectively extends several tens of thousands of kilometres into space, is called the magnetosphere.

MAGNETIC POLES

Two different types of magnetic poles must be distinguished. There are the "magnetic poles" and the "geomagnetic poles". The magnetic poles are the two positions on the Earth's surface where the magnetic field is entirely vertical. Another way of saying this is that the inclination of the Earth's field is 90° at the North Magnetic Pole and -90° at the South Magnetic Pole. A typical compass that is allowed to swing only in the horizontal plane will point in random directions at either the South or North Magnetic Poles.

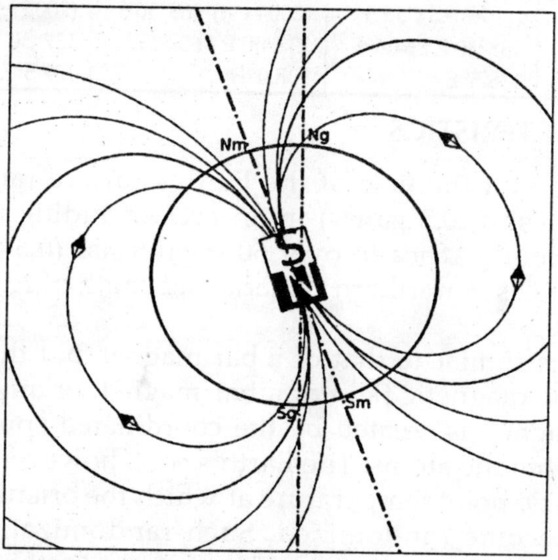

Fig. The Variation Between Magnetic North and "True" North

The Earth's field is closely approximated by the field of a dipole positioned near the centre of the Earth. A dipole defines an axis. The two positions where the axis of the dipole that best fits the Earth's field intersect the Earth's surface are called the North and South geomagnetic poles. If the Earth's field were perfectly dipolar, the geomagnetic and magnetic poles would coincide. However, there are significant non-dipolar terms which cause the position of the two types of poles to be in different places.

The locations of the magnetic poles are not static; they wander as much as 15 km every year. The pole position is usually not that which is indicated on many charts. The Geomagnetic Pole positions are usually not close to the position that commercial cartographers place

"Magnetic Poles." "Geomagnetic Dipole Poles", "IGRF Model Dip Poles", and "Magnetic Dip Poles" are variously used to denote the magnetic poles.

The Earth's field changes in strength and position. The two poles wander independently of each other and are not at directly opposite positions on the globe. Currently the magnetic south pole is farther from the geographic south pole than the magnetic north pole is from the geographic north pole.

Magnetic pole positions			
North Magnetic Pole	(2001) •81°18'N 110°48'W / 81.3°N 110.8°W	(2004 est) •82°18'N 113°24'W / 82.3°N 113.4°W	(2005 est) •82°42'N 114°24'W / 82.7°N 114.4°W
South Magnetic Pole	(1998) •64°36'S 138°30'E / 64.6°S 138.5°E	(2004 est) •63°30'S 138°00'E / 63.5°S 138.0°E	(2005 est) •63°06'S 137°30'E / 63.1°S 137.5°E

FIELD CHARACTERISTICS

The strength of the field at the Earth's surface ranges from less than 30 microteslas (0.3 gauss) in an area including most of South America and South Africa to over 60 microteslas (0.6 gauss) around the magnetic poles in northern Canada and south of Australia, and in part of Siberia.

The field is similar to that of a bar magnet, but this similarity is superficial. The magnetic field of a bar magnet, or any other type of permanent magnet, is created by the coordinated spins of electrons and nuclei within the atoms. The Earth's core, however, is hotter than 1043 K, the Curie point temperature at which the orientations of spins within iron become randomized. Such randomization causes the substance to lose its magnetic field. Therefore the Earth's magnetic field is caused not by magnetized iron deposits, but mostly by electric currents in the liquid outer core.

Convection of molten iron within the outer liquid core, along with a Coriolis effect caused by the overall planetary rotation, tends to organize these "electric currents" in rolls aligned along the north-south polar axis. When conducting fluid flows across an existing magnetic field, electric currents are induced, which in turn creates another magnetic field. When this magnetic field reinforces the original magnetic field, a dynamo is created which sustains itself. This is called the Dynamo Theory and it explains how the earth's magnetic field is sustained.

Another feature that distinguishes the Earth magnetically from a bar magnet is its magnetosphere. At large distances from the planet, this dominates the surface magnetic field. Electric currents induced in the ionosphere also generate magnetic fields. Such a field is always generated near where the atmosphere is closest to the Sun, causing daily alterations which can deflect surface magnetic fields by as much as one degree.

MAGNETIC FIELD VARIATIONS

Magnetometers detect minute deviations in the Earth's magnetic field caused by iron artifacts, kilns, some types of stone structures, and even ditches and middens in archaeological geophysics. Using magnetic instruments adapted from airborne magnetic anomaly detectors developed during World War II to detect submarines, the magnetic variations across the ocean floor have been mapped.

The basalt — the iron-rich, volcanic rock making up the ocean floor — contains a strongly magnetic mineral (magnetite) and can locally distort compass readings. The distortion was recognized by Icelandic mariners as early as the late 18th century. More important, because the presence of magnetite gives the basalt measurable magnetic properties, these magnetic variations have provided another means to study the deep ocean floor. When newly formed rock cools, such magnetic materials record the Earth's magnetic field.

Frequently, the Earth's magnetosphere is hit by solar flares causing geomagnetic storms, provoking displays of aurorae. The short-term instability of the magnetic field is measured with the K-index.

MAGNETIC FIELD REVERSALS

Based upon the study of lava flows of basalt throughout the world, it has been proposed that the Earth's magnetic field reverses at intervals, ranging from tens of thousands to many millions of years, with an average interval of approximately 250,000 years. The last such event, called the Brunhes-Matuyama reversal, is theorized to have occurred some 780,000 years ago.

There is no clear theory as to how the geomagnetic reversals might have occurred. Some scientists have produced models for the core of the Earth wherein the magnetic field is only quasi-stable and the poles can spontaneously migrate from one orientation to the other over the course of a few hundred to a few thousand years. Other scientists propose that the geodynamo first turns itself off, either spontaneously or through some external action like a comet impact, and then restarts itself with the magnetic "North" pole pointing either North or South. External events are not likely to be routine causes of magnetic field reversals due to the lack of a correlation between the age of impact craters and the timing of reversals. Regardless of the cause, when magnetic "North" reappears in the opposite direction this is a reversal, whereas turning off and returning in the same direction is called a geomagnetic excursion.

Studies of lava flows on Steens Mountain, Oregon, indicate that the magnetic field could have shifted at a rate of up to 6 degrees per day at some time in Earth's history, which significantly challenges the popular understanding of how the Earth's magnetic field works.

This has been found to be consistent, by measuring magnetism across ocean ridges. The molten lava (typically basalt or tholeiite) is extruded from volcanoes at well over the Curie temperature and then cools to adopt whatever magnetic field was present at the time. As time goes on more lava flows and bands of opposite magnetic fields are made present.

Using a magnetic detector (a variant of a compass), scientists have measured the historical direction of the Earth's magnetic field, by studying sequences of relatively iron-rich lava flows. Typically such layers have been found to record the direction of Earth's magnetic field when they cool. They have found that the poles have shifted a number of times throughout the past.

An alternate theory of the core of the Earth and the origin of the Earth's magnetic field has been proposed by J. Marvin Herndon. This proposal, known as the georeactor is not accepted by the scientific community.

GEOMAGNETIC REVERSAL

A geomagnetic reversal is a change in the orientation of Earth's magnetic field such that the positions of magnetic north and magnetic south become interchanged. These events often involve an extended decline in field strength followed by a rapid recovery after the new orientation has been established. These events occur on a scale of thousands of years or longer.

In the early 20th century geologists first noticed that some volcanic rocks were magnetized in a direction opposite to what was expected.

The first examination of the timing of magnetic reversals was done by Motonori Matuyama in the 1920s, who observed that there were rocks in Japan whose magnetic fields were reversed and those were all of early Pleistocene age or older. At the time he published his proposal suggesting that the magnetic field had been reversed, the magnetic field itself was poorly understood so there was little interest in the possibility that it had reversed.

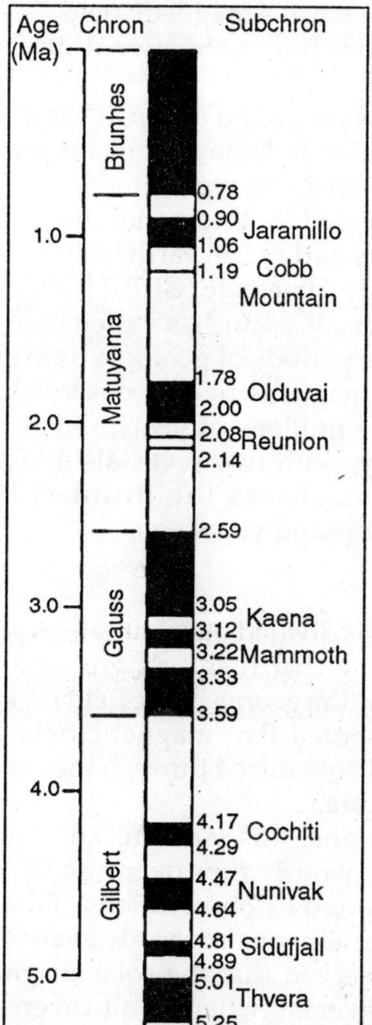

Three decades later, theories existed of the cause of the magnetic field and some of these included the possibility of field reversal. Most paleomagnetic research in the late 1950s was examining the wandering of the poles and continental drift. Although it was discovered that some rocks would reverse their magnetic field while cooling, it became apparent that most magnetized volcanic rocks contained traces of the Earth's magnetic field at the time the rock cooled. At first it seemed that reversals happen every one million years, but during the 1960s it became apparent that the time between reversals is erratic.

During the 1950s and 1960s research ships gathered information about variations in the Earth's magnetic field. Because of the complex routes of cruises, associating navigational data with magnetometer readings was difficult.

But when data was plotted on a map, it became apparent that there were remarkably regular and continuous magnetic stripes across the ocean floors. Through analysis of palaeomagnetic data, we now know that the field has reversed its orientation tens of thousands of times since its formation very early on in earth history. With the increasingly accurate Global Polarity Timescale (GPTS) it has become apparent that the rate at which reversals occur has varied considerably throughout the past. During some periods of geologic time (e.g. Cretaceous Long Normal), the Earth's magnetic field is observed to maintain a single orientation for tens of millions of years. Other events seem to have occurred very rapidly, with two reversals in a span of 50 thousand years. The last reversal was the Brunhes-Matuyama reversal approximately 780 thousand years ago.

Causes

Scientific opinion is divided on what causes geomagnetic reversals. Many scientists believe that reversals are an inherent aspect of the dynamo theory of how the geomagnetic field is generated. In computer simulations, it is observed that magnetic field lines can sometimes become tangled and disorganized through the chaotic motions of liquid metal in the Earth's core.

In some simulations, this leads to an instability in which the magnetic field spontaneously flips over into the opposite orientation. This scenario is supported by observations of the solar magnetic field, which undergoes spontaneous reversals every 7-15 years. However, with the sun it is observed that the solar magnetic intensity greatly increases during a reversal, whereas all reversals on Earth seem to occur during periods of low field strength.

Present computational methods have used very strong simplifications in order to produce models that run to acceptable time scales for research programs. A minority opinion, held by such figures as Richard A. Muller, is that geomagnetic reversals are not spontaneous processes but rather triggered by external events which directly disrupt the flow in the Earth's core. Such processes may include the arrival of continental slabs carried down into the mantle by the action of plate tectonics at subduction zones, the initiation of new mantle plumes from the core-mantle boundary, and possibly mantle-core shear forces resulting from very large impact events.

Supporters of this theory hold that any of these events could lead to a large scale disruption of the dynamo, effectively turning off the geomagnetic field. Because the magnetic field is stable in either the present North-South orientation or a reversed orientation, they propose that when the field recovers from such a disruption it spontaneously chooses one or the other state, such that a recovery is seen as a reversal in about half of all cases. Brief disruptions which do not result in reversal are also known and are called geomagnetic excursions.

OBSERVING PAST FIELDS

Past field reversals can be and have been recorded in the "frozen" ferromagnetic (or more accurately, ferrimagnetic) minerals of solidified sedimentary deposits or cooled volcanic flows on land. Originally, however, the past record of geomagnetic reversals was first noticed by observing the magnetic stripe "anomalies" on the ocean floor. Lawrence W. Morley, Frederick John Vine and Drummond Hoyle Matthews made the connection to seafloor spreading in the Morley-Vine-Matthews hypothesis which soon led to the development of the theory of plate tectonics.

Given that the sea floor spreads at a relatively constant rate, this results in broadly evident substrate "stripes" from which the past magnetic field polarity can be inferred by looking at the data gathered from towing a magnetometer along the sea floor. However, because no existing unsubducted sea floor (or sea floor thrust onto continental plates, such as in the case of ophiolites) is much older than about 180 million years (Ma) in age, other methods are necessary for detecting older reversals. Most sedimentary rocks incorporate tiny amounts of iron rich minerals, whose orientation is influenced by the ambient magnetic field at the time at which they formed. Under favorable conditions, it is thus possible to extract information of the variations in magnetic field from many kinds of sedimentary rocks. However, subsequent diagenetic processes after burial may erase evidence of the original field.

Because the magnetic field is present globally, finding similar patterns of magnetic variations at different sites is one method used to correlate age across different locations. In the past four decades great amounts of paleomagnetic data about seafloor ages (up to ~250 Ma) have been collected and have become an important and convenient tool to estimate the age of geologic sections. It is not an independent dating method, but is dependent on "absolute" age dating methods like radioisotopic systems to derive numeric ages. It has become especially useful to metamorphic and igneous geologists where the use of index fossils to estimate ages is seldom available.

THE GEOMAGNETIC POLARITY TIME SCALE

GEOMAGNETIC REVERSALS

The rate of reversals in the Earth's magnetic field has varied widely over time. 72 million years ago (Ma), the field reversed 5 times in a million years. In a 4-million-year period centered on 54 Ma, there were 10 reversals; at around 42 Ma, 17 reversals took place in the span of 3 million years. In a period of 3 million years centering on 24 Ma, 13 reversals occurred. No fewer than 51 reversals occurred in a 12-million-year period, centering on 15 million years ago. These eras of frequent reversals have been counterbalanced by a few "superchrons" – long periods when no reversals took place.

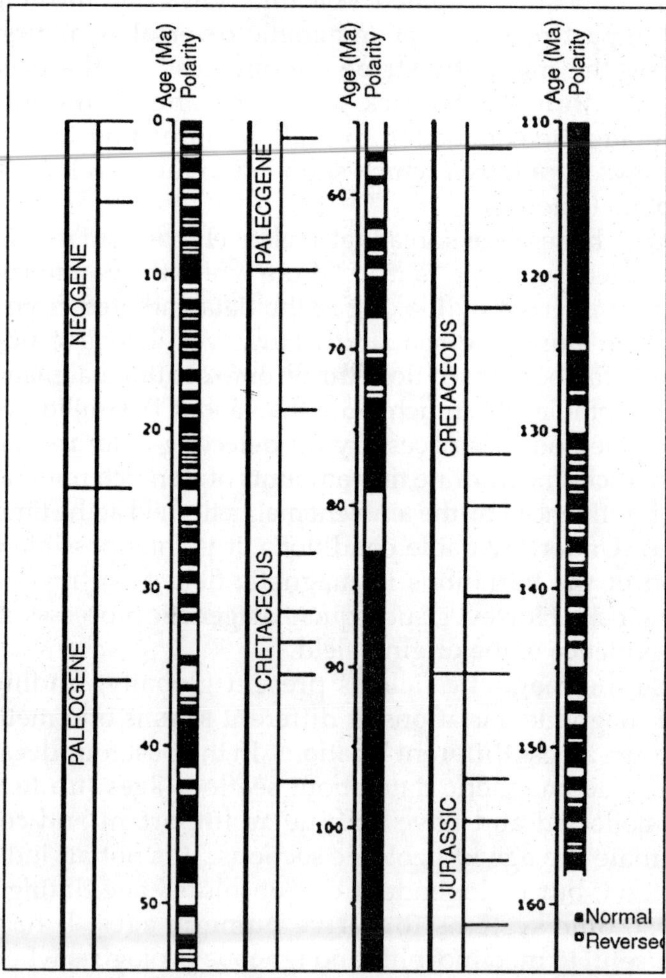

Fig. Record of Geomagnetic Polarity for Ages 0-160 Ma.

It had generally been assumed that the frequency of geomagnetic reversals is random, and it was shown in 2006 that the known reversals conform to a Lévy distribution.

THE CRETACEOUS LONG NORMAL SUPERCHRON

A long period of time during which there were no magnetic pole reversals, the Cretaceous Long Normal (also called the Cretaceous Superchron or C34) lasted from about 120 to 83 million years ago. This time period included stages of the Cretaceous period from the Aptian through the Santonian. An interesting trend can be seen when looking at the frequency of magnetic reversals approaching and following the Cretaceous Long Normal. The frequency steadily decreased prior to the period, reaching its low point (no reversals) during the period. Following the Cretaceous Superchron the frequency of reversals slowly increased over the next 80 million years, to the present.

The Jurassic Quiet Zone

The Jurassic Quiet Zone is a section of ocean floor which is completely devoid of the magnetic stripes that can be detected elsewhere. This could mean that there was a long period of polar stability during the Jurassic period similar to the Cretaceous Superchron. Another possibility is that as this is the oldest section of ocean floor, any magnetization that did exist has completely degraded by now. The Jurassic Quiet Zones exist in places along the continental margins of the Atlantic ocean as well as in parts the Western Pacific.

The Kiaman Long Reversed Superchron

This long period without geomagnetic reversals lasted from approximately the late Carboniferous to the late Permian, from approximately 316 to 262 million years ago. The magnetic field was reversed compared to its present state. The name "Kiaman" derives from the Australian village of Kiama, where some of the first geological evidence of the superchron was found in 1925.

The Moyero Reversed Superchron

This period in the Ordovician (485 to 463 million years ago) is suspected to host another superchron. But until now this possible superchron was only found in the Moyero river section north of the polar circle in Siberia.

FUTURE OF THE PRESENT FIELD

At present, the overall geomagnetic field is becoming weaker at a rate which would, if it continues, cause the dipole field to temporarily collapse by 3000–4000 CE. The South Atlantic Anomaly is believed by some to be a product of this. The present strong deterioration corresponds to a 10–15% decline over the last 150 years and has accelerated in the past several years; however, geomagnetic intensity has declined almost continuously from a maximum 35% above the modern value achieved approximately 2000 years ago. The rate of decrease and the current strength are within the normal range of variation, as shown by the record of past magnetic fields recorded in rocks.

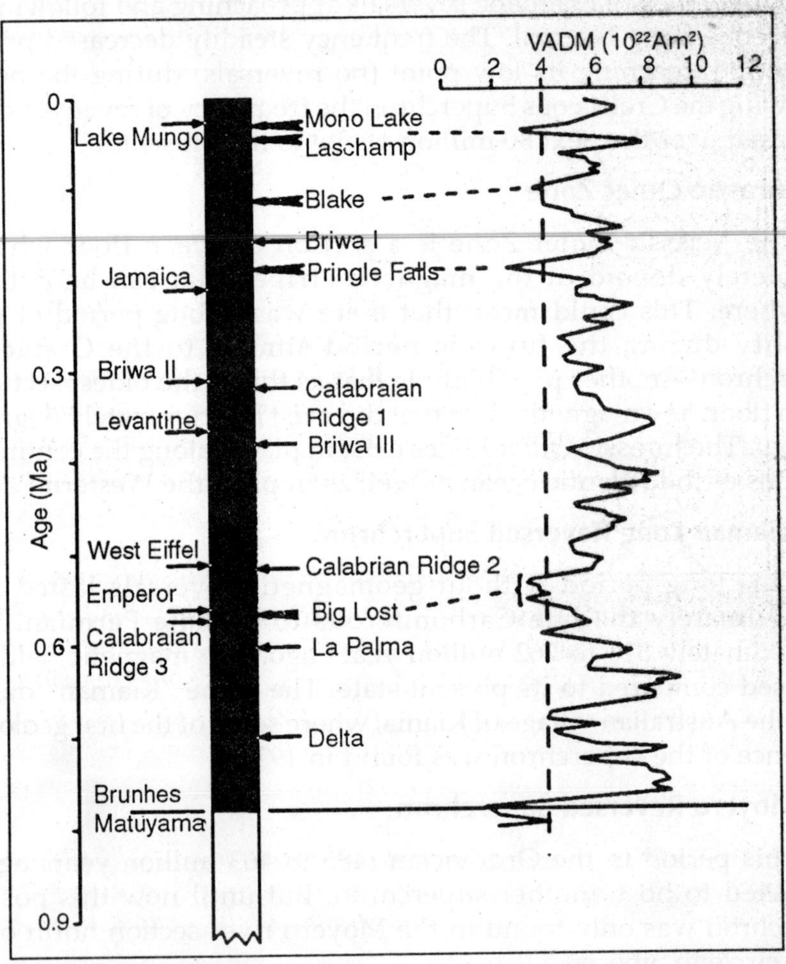

Fig. Geomagnetic Variations Since the Last Reversal

The nature of Earth's magnetic field is one of heteroscedastic fluctuation. An instantaneous measurement of it, or several measurements of it across the span of decades or centuries, is not sufficient to extrapolate an overall trend in the field strength. It has gone up and down in the past with no apparent reason. Also, noting the local intensity of the dipole field (or its fluctuation) is insufficient to characterize Earth's magnetic field as a whole, as it is not strictly a dipole field. The dipole component of Earth's field can diminish even while the total magnetic field remains the same or increases.

The Earth's magnetic north pole is drifting from northern Canada towards Siberia with a presently accelerating rate — 10km per year at the beginning of the 20th century, up to 40km per year in 2003. It is also unknown if this drift will continue to accelerate.

Glatzmaier and collaborator Paul Roberts of UCLA have made a numerical model of the electromagnetic, fluid dynamical processes of Earth's interior, and computed it on a Cray supercomputer. The results reproduced key features of the magnetic field over more than 40,000 years of simulated time. Additionally, the computer-generated field reversed itself

EFFECTS ON BIOSPHERE AND HUMAN SOCIETY

Because the magnetic field has never been observed to reverse by humans with instrumentation, and the mechanism of field generation is not well understood, it is difficult to say what the characteristics of the magnetic field might be leading up to such a reversal. Some speculate that a greatly diminished magnetic field during a reversal period will expose the surface of the earth to a substantial and potentially damaging increase in cosmic radiation. However, Homo erectus and their ancestors certainly survived many previous reversals. There is no uncontested evidence that a magnetic field reversal has ever caused any biological extinctions. A possible explanation is that the solar wind may induce a sufficient magnetic field in the Earth's ionosphere to shield the surface from energetic particles even in the absence of the Earth's normal magnetic field.

Although the inspection of past reversals does not indicate biological extinctions, present society with its reliance on electricity and electromagnetic effects (e.g. radio, satellite communications) may be vulnerable to technological disruptions in the event of a full field reversal.

Chapter 5

Magnetic Properties of Soil

One of the main problems of using modern technology for the clearance of buried land mines and unexploded ordnance (UXO) is the large probability of false alarms and the associated time and labour that is required to excavate all suspect objects. Most soil physical properties exhibit significant variability at a wide range of scales, which poses a problem for many sensors that are used for the detection of buried objects. The natural variation in the soil properties gives rise to anomalies, similar to anomalies that are the result of buried objects.

The discrimination between these geological or soil anomalies and buried object anomalies is extremely complicated and is one of the major problems that need to be addressed in order to improve the performance of sensors for buried land mine and UXO detection. Geophysical instruments for buried object detection that use electromagnetic waves are various: time- and frequency domain EM, ground penetrating radar, and magnetic methods. However, the performance of these instruments can degrade significantly at sites with variable soil properties such as water content, texture, and magnetic characteristics.

Several studies in the past have addressed the influence of variability in soil texture and soil water content on electromagnetic sensors for the detection of land mines. The study of magnetic soil properties in relation to the problem of land mine and UXO detection is relatively new. Also, little is known about the natural variability in magnetic characteristics of soils.1 In strongly magnetic soils, magnetic and electromagnetic sensors often detect anomalies that have a geologic or pedogenic origin.

To improve the discrimination performance of land mine and UXO detection sensors it is necessary to get a better understanding of the magnetic variability at survey sites, and to use the statistical information on the distribution of magnetic soil properties in data processing algorithms. To address this issue, we asked ourselves three basic questions:

- What causes magnetic soil properties,
- What is the spatial variability in magnetic soil properties at different locations,
- How can we explain the variability within a site and between different sites?

MAGNETICS

The physical background for the existence of magnetic behaviour in minerals is the magnetic moment produced by electrons orbiting their nucleus and spinning around their axis. In many types of material the overall magnetic moment is zero because the orbital and spin components even out. When a mineral with zero magnetic moment is placed in a magnetic field the electron motions will rearrange so that the net magnetic moment is in the direction opposite to the applied field.

These types of minerals are called *diamagnetic*. In contrast, when minerals with a small net magnetic moment get subjected to a magnetic field the electrons will attempt to line up in the direction of the magnetic field. These types of minerals are called *paramagnetic*. In some minerals, the interaction between electron spin and orbital movement in adjacent atoms causes these minerals to behave as active magnets.

These types of minerals are called *ferromagnetic* when all magnetic moments line up in the same direction, or *ferrimagnetic*, when one-third of the magnetic moments line up in the opposite direction. A special group of minerals are those in which the electron interaction leads to magnetic moments being aligned in opposite directions. These minerals with a net magnetic moment of zero are called *antiferromagnetic*.

MAGNETIC SOIL PROPERTIES

Magnetic properties in soils are largely a consequence of the presence of different forms of iron. Although pure iron can occur naturally in rocks and soil, it is very rare. Specific types of iron oxides, iron-titanium oxides and iron sulfides are the predominant causes of magnetic soil characteristics. In abundance, iron (Fe) is the fourth element in the earth crust. Although the most abundant minerals in the earth's continental crust are essentially Fe-free (plagioclase, feldspar, quartz), many other minerals contain significant amounts of iron.

Iron-containing minerals can be found in igneous rock such as basalt, gabbro, and granite, but also in metamorphic and sedimentary rocks. Therefore, it is no surprise that iron occurs in one form or another in many soils. The concentration of (magnetic) iron oxides is affected by the parent material, soil age, soil forming processes, biological activity, and soil temperature. Table shows magnetic susceptibilities for several iron- and iron-titanium-oxides, iron-sulfides and other soil constituents. Water and quartz are diamagnetic and have a small negative magnetic susceptibility.

Hydrated iron oxides like goethite, which is the most abundant iron oxide in soils around the world, ferrihydrite, and lepidocrocite, play a minor role in determining the magnetic character of soils. Also hematite, which is the most abundant iron oxide in tropical soils, pyrite, and ilmenite hardly affect the magnetic soil characteristics. The magnetic character of soils is dominated by the presence of ferrimagnetic minerals such as magnetite and maghemite, and to a lesser degree by pyrrhotite.

Although iron oxides occur in most environments throughout the world 18 some locations are more favorable for the formation and/or maintenance of significant amounts of (magnetic) iron oxides than others. Unfortunately, soil maps and available laboratory data usually only contain information on the amount and not the type of iron oxides 19. Tropical soils often contain large amounts of iron oxides 20. Many tropical soils have deeply weathered profiles whose red and yellow colors result from an accumulation of iron and aluminum oxides.

Table.Magnetic Susceptibilities for Several Iron Oxides and Soil Constituents

Material	Chemical Formula	Magnetic Status	Magnetic Susceptibility (10^{-8} M^3 kg^{-1})
Water	H_2O	Diamagnetic	-0.9
Quartz	SiO	Diamagnetic	-0.6
Pyrite	FeS_2	Paramagnetic	30
Ferrihy drite	$5Fe_2O_3.9H_2O$	Paramagnetic	40
Leopido crocite	γ -FeO.OH	Paramagnetic	70
Ilmenite	$FeTiO_3$	Superparamagnetic	200
Hematite	α -Fe_2O_3	Antiferromagnetic	60
Goethite	α -FeO.OH	Antiferromagnetic	70
Pyrrhotite	$Fe_7S_8/Fe_8S_9/Fe_9S_{10}$	Ferrimagnetic	~5,000
Maghemite	γ -Fe_2O_3	Ferrimagnetic	40,000
Magnetite	Fe_3O_4	Ferrimagnetic	50,000

Large areas of these soils can be found in Africa and South America with minor acreages in South-East Asia. Another type of soil where iron oxides are abundant are relatively young soils developed from parent material of volcanic origin. The volcanic origin ensures in many cases the abundance of (ferri)magnetic iron oxides. Volcanic derived soils have a much smaller acreage than Fe-rich soils in the tropics but can be found on all continents. They may be present everywhere where geologically young volcanic rock is found (e.g., near continental margins and subduction zones).

CHARACTERIZATION OF MAGNETIC SOIL PROPERTIES

There are three magnetic effects that impact the (electro)magnetic characteristics of the subsurface, and thus electromagnetic sensors:

- Remanent magnetization,
- Induced magnetization,
- Viscous remanent magnetization.

Remanent Magnetization

Remanent magnetization exists in the absence of an applied field. The remanent magnetization must be added to any magnetization effects resulting from an applied magnetic field. Remanent magnetization occurs within ferromagnetic and ferromagnetic minerals that have a natural alignment of the magnetic moments. This type of magnetization directly affects magnetic sensors. Also, remanent magnetization can be the result of alignment and subsequent 'locking' of magnetic moments in the weak magnetic field of the Earth. Locking can occur due to cooling from high temperatures through the mineral-specific Curie temperature, due to critical-size crystal growth, or due to compaction and consolidation. These thermoremanent, chemical-remanent, and detrital-remanent magnetizations are small and affect only the most sensitive magnetic sensors.

Induced Magnetization

Induced magnetization results from a magnetic field being applied to a magnetically susceptible object. In the low-intensity field region, the net magnetic moment (i.e., the magnetization, *M*) is proportional to the strength of the applied field (*H*). Therefore, the low-field magnetic susceptibility, defined as the ratio of the magnetization over the field strength, is a material-specific property. The magnetic susceptibility is either expressed per unit volume (volume-specific susceptibility,) or per unit mass (mass-specific susceptibility, Ç). Induced magnetization can be measured by applying a magnetic field to a sample (in the laboratory or in the field).

By measuring the difference between this primary magnetic field and the secondary magnetic field one can determine the material specific magnetic susceptibility. The magnetic induction of a sample, measured by a magnetic or electromagnetic sensor, is the sum of all the different entities of induced magnetization, weighed for volume, distance to the sensor, and magnitude of the susceptibility. Ferrimagnets are the most important minerals for affecting the magnetic susceptibility.

VISCOUS REMANENT MAGNETIZATION

Viscous remanent magnetization refers to the effect that the secondary magnetic field gets delayed relative to the primary magnetic field 11. This effect differs from the standard induced magnetization, where the magnetization is instantaneous, and the secondary magnetic field is in-phase with the primary magnetic field. Viscous remanent magnetization occurs in ferrimagnetic materials with a range of different shapes (anisotropy) and especially grain sizes. Under these circumstances, the application or removal of a magnetic field to/from a sample causes a delayed change in the direction of magnetization. The time needed for the direction change to occur is known as the Neel relaxation time.

One consequence of viscous remanent magnetization is that the susceptibility becomes frequency dependent. This effect has important implications for both time- and frequency-domain electromagnetic sensors. Viscous remanent magnetization can be measured using dual-frequency magnetic susceptibility sensors. To study the variability of magnetic soil characteristics we have selected a number of sites in the tropics, for they are expected to have significant amounts of iron oxides: Hawaii, Panama, and Ghana. Furthermore we have gathered information from two locations in the continental United Sates, in New Mexico and Wyoming. In our magnetic measurements we have focused on induced magnetization (magnetic susceptibility) and viscous remanent magnetization (frequency dependent magnetic susceptibility). For these measurements we used Bartington magnetic equipment. We have not separately measured the remanent magnetization of the material.

Table.Overview of Bartington for Measurement of Magnetic Susceptibility

Bartington sensor	MS2D	MS2F	MS2B
Application	Field	Field	Lab
Frequency	0.96 kHz	0.58kHz	0.46/4.6kHz
Area/Volume	268.7 cm^2	1.8 cm^2	10 cm^3

At most of the sites we have selected one or more transects in agricultural fields for in-situ measurement of magnetic susceptibility at the surface. Also, we have collected sample material for further analysis in the laboratory. The sampling locations were partly based on the extremes in magnetic susceptibility readings, and partly using a stratified random sampling scheme where we took one random sample from every 5 meter interval. In many of the locations we were able to use soil pits in order to study soil development and to measure the variation in magnetic properties with depth. From these soil pits we have taken samples for analysis in the laboratory.

In the laboratory we performed a wide range of tests, such as frequency dependent magnetic susceptibility measurements (using the Bartington MS2B sensor). Also, we have measured textural characteristics (bulk density and grain size analysis), and we have performed a number of tests (X-ray fluorescence spectroscopy, magnetic iron extraction) to better understand the type and quantity of iron oxides. At all the studied locations on the Big Island of Hawaii the soils were formed in similar parent material (volcanic ashes) with comparable ages of 20k to 30k years. This allowed us to study the role of climate on soil forming processes in general, and magnetic soil properties in particular.

Our assumption is that the degree of soil formation is a result of, amongst others, soil age and climatic factors. Due to the position in the trade wind belt and the general shape of the island (high mountains with steep slopes) the Island of Hawaii experiences extreme gradients in rainfall 24,25. The windward eastern slopes of the island receive larger amounts of rainfall than the leeward western slopes, while on both sides the rainfall increases with elevation. The mean annual rainfall is greatest at around 1500 meters elevation and then decreases with further elevation, which is a result of the air flowing around rather than over the highest mountains. We have selected 7 locations for our measurements; 3 on Kohala Volcano, and 4 on Mauna Kea Volcano. Lines 1, 2, and 3 gradually decrease in elevation and mean annual rainfall and lie on an imaginary line on the leeward side of Kohala Volcano. The same holds for Lines 5, 7 and 6 on the windward side of Mauna Kea. Line 4 lies on the other side of Mauna Kea and experiences a much lower mean annual rainfall.

SPATIAL VARIABILITY

The spatial variability in magnetic susceptibility was measured along several transects on the Big Island of Hawaii and the Island of O'ahu, as well as in Panama, and New Mexico. There exists considerable variability at small scales and between the sites. It can be seen in the scatter plots that for most sites the minimum and maximum readings differ by a factor between 2 to 5.

O'AHU, HAWAII

Figure shows the variability in magnetic susceptibility for Shofield Barracks on O'ahu, Hawaii, measured with two types of sensors. The difference between the readings of the MS2D and MS2F sensor readings are a result of differences in sensitivity 26. Also, the sensors relate to different volumes of material, and because soils are rarely homogeneous, different sensors may result in different readings. Furthermore, the field measurements depend highly on the ground coupling of the probe. Air gaps will significantly decrease the measured susceptibilities; the MS2F probe is most vulnerable to this problem.

The data in Figure which show densely spaced measurements along the same line as in Figure show that the small MS2F sensor is not able to pick up a lot more variation than is observed with the larger-diameter MS2D sensor. Also, the MS2F sensor gives significantly lower susceptibility readings. These considerations have made us decide to do all further measurements with the MS2D rather than the MS2F sensor.

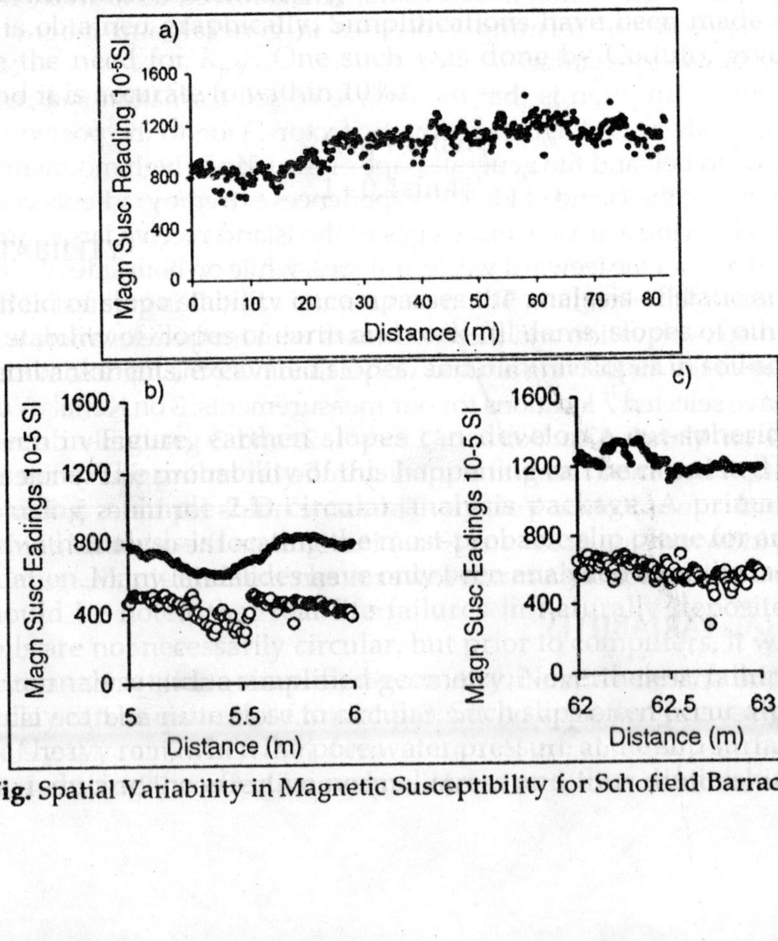

Fig. Spatial Variability in Magnetic Susceptibility for Schofield Barracks

PANAMA

Transects 1 and 3 for Panama show magnetic susceptibility readings with a somewhat larger spatial variability than in O'ahu, Hawaii. The one notable exception relative to all other sites with respect to absolute susceptibility readings and variability within a site is Line 4 in Panama. This site was the only one not in agricultural land, but on a beach, rich in pure magnetite. Here, the continuous hydraulic sorting in the surf zone has caused the heavy minerals (magnetite) and the quartz sand to be deposited in different areas. Due to the large extremes and the higher than average variability many sensors for the detection of land mines and UXO will experience significant problems.

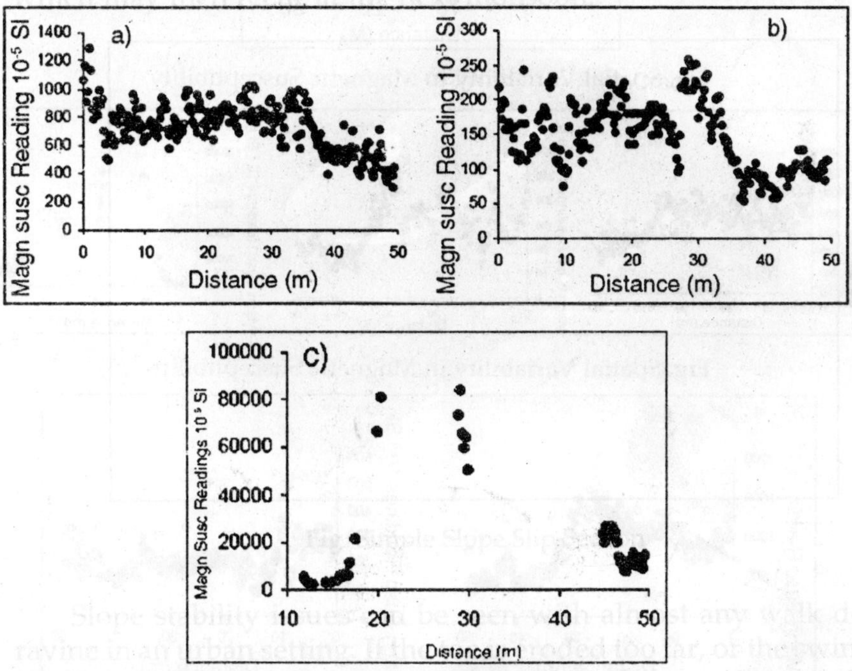

Fig. Spatial Variability in Magnetic Susceptibility

For reference we collected a transect in Socorro, New Mexico. The soil material is wind blown sand, or loess. The climate can be described as semi arid, which results in totally different soils than for the rest of the transects presented in this paper.

Many soils in New Mexico are characterized by the presence of so-called calcic horizons within the first meter of the soil profile. These calcic horizons are accumulations of salts that form either through precipitation or biochemical alterations. The susceptibility readings along a 35m long transect show a variability that is comparable to the other locations, and absolute numbers that are comparable with low magnetic susceptibility sites in the tropics such as in Panama.

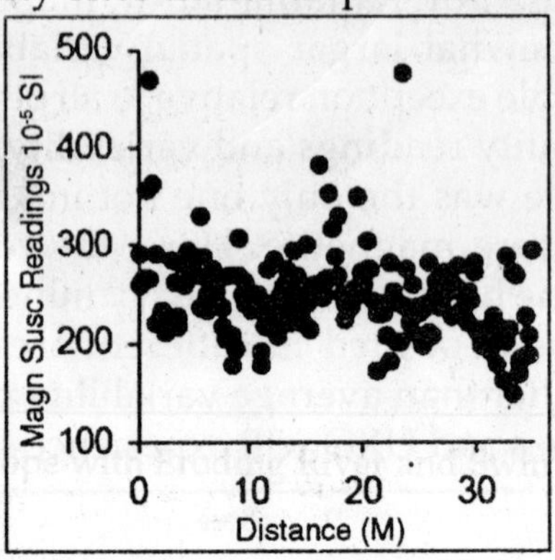

Fig. Spatial Variability in Magnetic Susceptibility

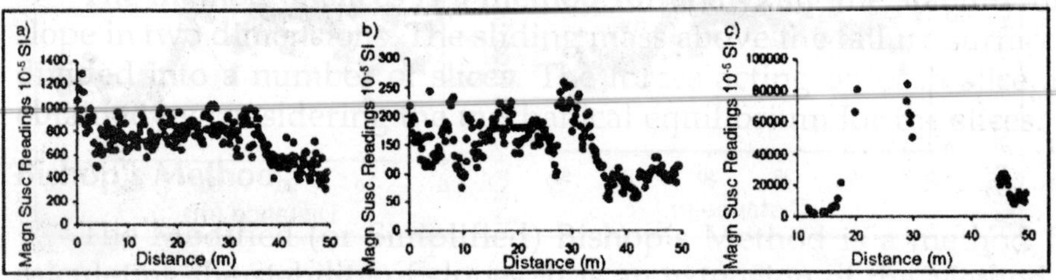

Fig. Spatial Variability in Magnetic Susceptibility

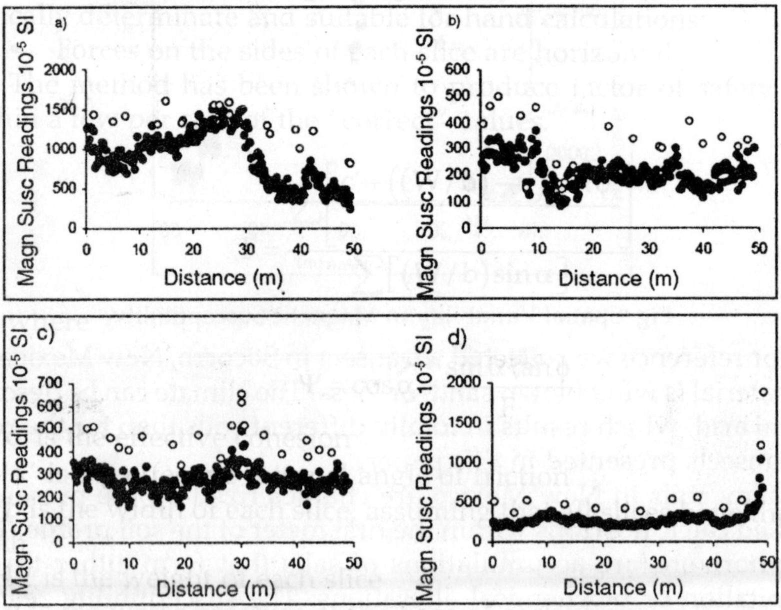

Fig. Spatial variability in Magnetic Susceptibility

BIG ISLAND

For all transects, the magnetic susceptibility readings are below 1000 10-5 SI. The spikes in the readings are the result of volcanic rocks or sinter blocks at or near the surface. Probably, in these rocks a larger amount of magnetite is present than in the surrounding soil. The sudden change in susceptibility readings at a distance of 30m on Line 4 follows the transition from a lava flow onto a soil formed in depositional material. Much higher magnetic susceptibilities were measured at Line 4 and especially Line 3 than for the rest of the transects. In the soil of Line 3, many black grains behaving as small magnets were found, indicating that significant amounts of magnetite are present in this soil. Considering the fact that all these soils have the same parent material, it seems that dry conditions favour the occurrence of ferrimagnetic minerals.

Nevertheless, it seems impossible to make a direct correlation between susceptibility readings and mean annual rainfall. For the transects 1 to 3 on Kohala Volcano, the average mean annual rainfall decreases from 1300 mm/y, via 750 mm/y to 180 mm/y while the average susceptibility readings are 325·10-5 SI, 198·10-5 SI, and 1375·10-5 SI, respectively. Part of this bad correlation can be attributed to the presence of vegetation, which has a strong effect on the susceptibility readings. The open circles in the diagrams in Figures represent measurements at the locations that were selected for sampling. Here, the vegetation was removed and before the sample material was collected we repeated the susceptibility measurements. The data Figures show that in many instances the vegetation has a strong effect on the measurements.

VERTICAL VARIABILITY IN MAGNETIC SUSCEPTIBILITY

Magnetic measurements in soil pits have three advantages over measurements at the surface. Firstly, soil pits allow for the study of soil development. This may provide clues to magnetic behaviour. Secondly, soil pits eliminate the effect that vegetation has on the measurements. Even when vegetation is removed, the presence of a litter layer can lower the surface readings. Thirdly, most EM sensors for the detection of buried land mines and UXO have a larger penetration depth than the highly sensitive Bartington sensors.

Due to the shallow penetration of the Bartington sensors, surface measurements can not fully characterize all the variability in magnetic soil properties. Figure shows the results of the magnetic measurements in the soil pits, which were done using a vertical sensor orientation. For half of the soil pits the magnetic susceptibility increases with depth, while for the other half a decrease with depth is observed. In all but one soil pit we observe a gradual change in magnetic susceptibility with depth. Only in soil pit 3 on the Big Island of Hawaii the readings experience a sudden large change. Future chemical analysis will have to explain this trend.

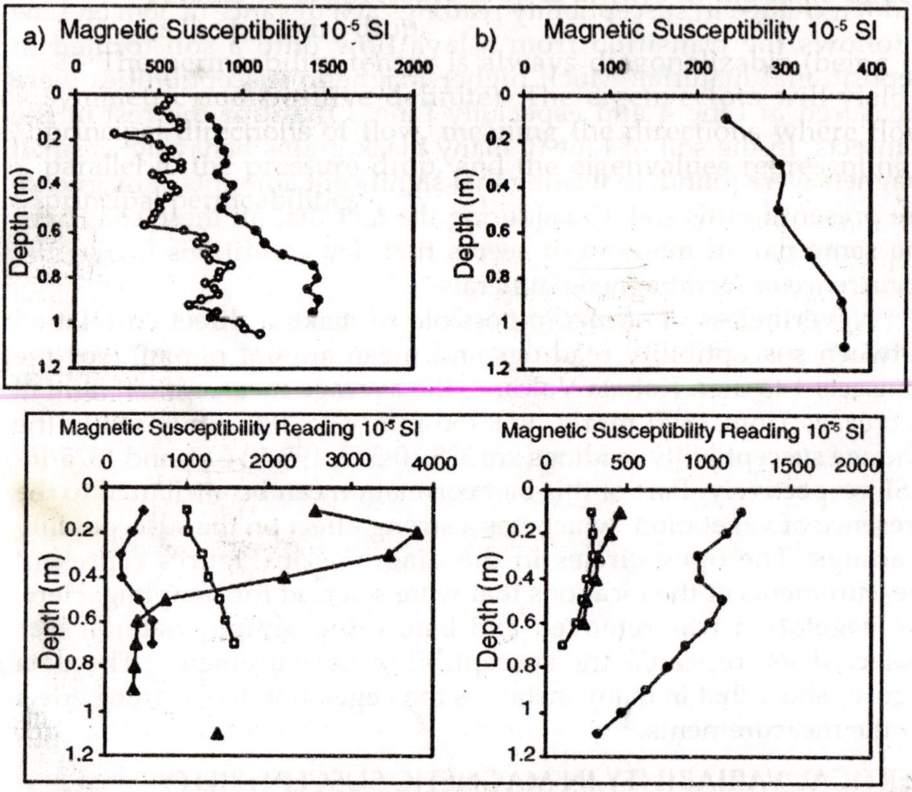

Fig. Vertical variation in Magnetic Susceptibility in Soil pits.

- O'ahu, Hawaii. The Open Circles Represent Measurements with the Bartington MS2F sensor while the black circles are results of the MS2D sensor measurements.
- Panama, Line 3.
- Kohala Volcano, Big Island, Hawaii. The black diamonds and open squares represent soil pits measurements for Lines 1 and 2, respectively, while the triangles represent the measurements in the soil pit at Line 3. The measurement at 1.1 m depth is for unweathered basalt. c) Mauna Kea Volcano, Big Island, Hawaii. The black circles, open squares, and the triangles represent measurements in soil pits at Lines 4, 5 and 6, respectively.

CORRELATION OF MEAN ANNUAL RAINFALL AND MAGNETIC SUSCEPTIBILITY

Comparison of the magnetic susceptibility readings and values of the mean annual rainfall at the Big Island of Hawaii suggests there may be correlation between the two. Figure shows average magnetic susceptibility readings for surface measurements and soil pits plotted versus the mean annual rainfall. The averages for the surface were calculated based on the measurements at the 20 sampling locations where vegetation was removed. The averages for the soil pits were calculated over the interval from 0.1 to 0.6 m.

This range is chosen somewhat arbitrarily but is the maximum range for which we were able to take measurements in all 6 soil pits. The results in Figure show a clear trend of decreasing magnetic susceptibility readings with an increasing mean annual rainfall. When combining the two datasets (i.e., averaging between the values from the soil pits and the surface measurements) a power function of the form y=22834x-0.5087 fits the data with an R2 of 0.963. Although a few assumptions and simplifications were made this strong fit demonstrates the clear correlation between mean annual rainfall and magnetic susceptibility.

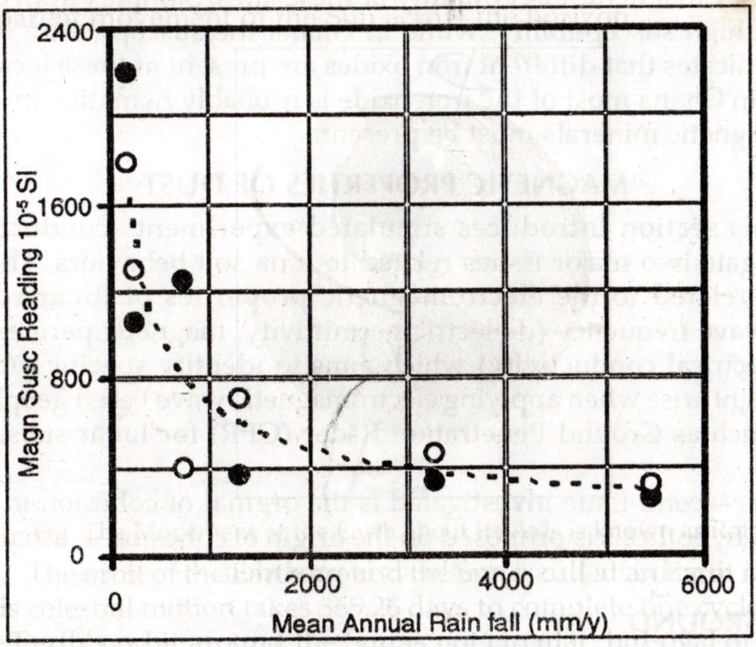

Fig. Cross Plot of Average Magnetic Susceptibility Readings Versus Mean Annual Rainfall for 6 Locations on the Big Island of Hawaii.

LABORATORY MEASUREMENTS

Laboratory analysis provides the opportunity to determine the magnetic minerals responsible for the measured magnetic susceptibility in soils. The most important question to be answered is the frequency dependence of the material.

When comparing the average low- and high frequency magnetic susceptibility readings with the frequency dependence one important observation can be made: there is no correlation between the absolute magnetic susceptibility and the frequency dependence of a sample. O'ahu and Panama both have a similar magnetic susceptibility, but while the material from Panama exhibits virtually no frequency dependence, the samples from O'ahu have a very large frequency dependence.

Also some samples from Ghana show significant frequency dependence, while for these samples the high- and low-frequency magnetic susceptibility appears to be very low. The total iron content (using XRF) was determined only for samples from O'ahu and Ghana. These data show that the samples from both these locations contain a significant percentage of iron.

However, there is a large difference between he high- and low-frequency magnetic susceptibility at these sites. Samples from O'ahu have a high susceptibility, while in Ghana the susceptibility is low. This indicates that different iron oxides are present at these locations. While in Ghana most of the iron oxide is probably hematite, in O'ahu ferrimagnetic minerals must be present.

MAGNETIC PROPERTIES OF DUST

This section introduces simulated experiments conducted to investigate two major issues related to luna soil behaviors. The first one is related to the electromagnetic properties of lunar soils at microwave frequency (dielectric permittivity, magnetic permeability and electrical conductivity) which aims to identify specific problem that might arise when applying electromagnetic wave based geophysics tools such as Ground Penetration Radar (GPR) for lunar subsurface survey.

The second issue investigated is the orginal of cohesion in lunar soils. A hypothesis is proposed on the origin of cohesion in lunar soils. Method to validate this hypothesis is provided.

BACKGROUND

Electromagnetic wave based technology such as ground penetration radar provides efficient tool for subsurface investigations.

This technology is based on the propagation and reflections of electromagnetic wave due to the electromagnetic properties of a material.

When apply electromagnetic wave technology on the earth, the most important component that need special attention is water. This is due to its large dielectric constant and high electrical conductivity. The amount of water has important effects on the speed, dispersion and attenuation of of electromagnetic wave propagating into the ground.

The magnetic permeability of geomaterials are typically assumed to be around one for GPR applications as most earth materials are diamagnetism or parmagnetism. The travel speed of an electromagnetic wave in a material, *V*, is related to its relative dielectric permittivity,

•γ and relative magnetic permeability, μγ byEq $V = \frac{c}{\sqrt{\varepsilon_r \mu_r}}$

by Eq. where c is the speed of light. Compare with the earth, lunar soil is absent of water. On the other hand, it features gluttinitic glass structure with nanophase metallic Fe grains.

FeO on average accounts for around 15% of the total weights. As ferromagnetic materials, FeO shows distinctive large ferromagnetic behaviors. The influence of FeO nanoparticles on the electromantic properties of lunar soils thus need to be investigated to properly interpret radar signals.

METHOD

Measurement of electromagnetic properties was conducted using Time Domain Reflectometry (TDR), which is a guided electromagnetic wave technology. A schematic test set up is shown in Figure. TDR utilizes the propagation of electromagnetic wave to measure material properties. It works by generating a small-magnitude electromagnetic field excitation and measures the material response.

By using a fast rising pulse of a few picoseconds by the current electronics, TDR measures the broad frequency band material responses from a few megahertz to gigahertz. The information commonly used from a TDR signal is the reflection points, which are related to the speed of electromagnetic wave in the soil and are subsequently used for determining the apparent dielectric constant *Ka*; and the long term signal level, which is related to the energy attenuation and subsequently used to determine the electrical conductivity *ECb*. Both quantities can be easily obtained from a TDR signal.

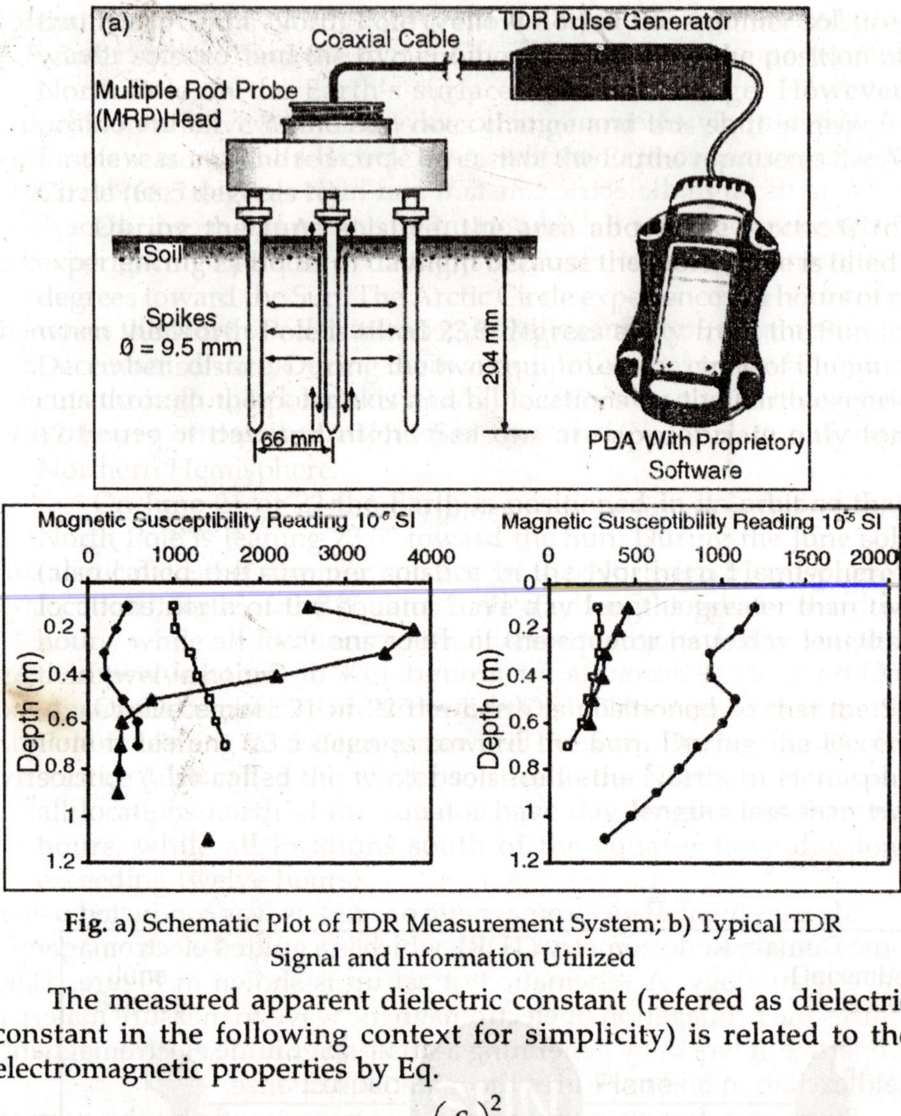

Fig. a) Schematic Plot of TDR Measurement System; b) Typical TDR Signal and Information Utilized

The measured apparent dielectric constant (refered as dielectric constant in the following context for simplicity) is related to the electromagnetic properties by Eq.

$$K_a = \left(\frac{c}{V}\right)^2 = \varepsilon_r \mu_r$$

In this experimental programme, tests were conducted in a cylinder mold with known volume. TDR test was first conducted on pure fine sand. FeO powder is weighted and mixed with sand at mass ratio of 5%, 10%, 15%, 30%.

Mixtures were placed into the mold and TDR tests were taken, from which the dielectric constant and electrical conductivity were determined. Based on the amount of mixture placed, the density and percentage of different phases were also determined. An example of measured signal is shown in Figure. The sharp rising at the end of probe indicates there is no appreciable dispersion occurs for the simulated lunar soil mixture. Although FeO is conductive, the electrical conductivity of whole mixture is low. This might be due to the fact that they have not formed a percolation channels to allow current to be conducted across the sample.

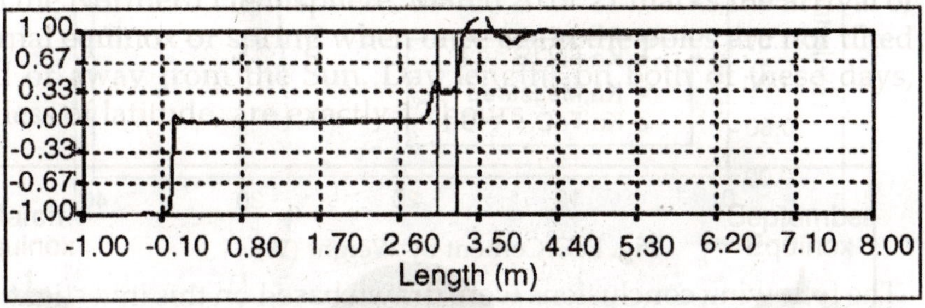

Fig. Example of TDR Measured Signal of Sand FeO Powder Mixture

Analyses

The dielectric constant measured by TDR is the bulk value of the mixture. It is related to the volumetric components by the following relationship:

$$K_a^n = n_{air} \cdot K_{a,air}^n + n_{sand} \cdot K_{a,sand}^n + n_{FeO} \cdot K_{a,FeO}^n$$

Where $n_{air}, n_{sand}, n_{FeO}$ are volume of air, sand and FeO respectively. *Ka* with appropriate subscript stands for their respective dielectric constant. n is an index related to particle orientation, n has a value of 0.5 when particles are randomly oriented. Substitute Equation into Equation there is:

$$\sqrt{K_a} = n_{air}\sqrt{\varepsilon_{r,air}\mu_{r,air}} \cdot + n_{sand} \cdot \sqrt{\varepsilon_{r,sand}\mu_{r,sand}} \cdot + n_{FeO} \cdot \sqrt{\varepsilon_{r,FeO}\mu_{r,FeO}} \cdot$$

where •*r* and μ*r* with respective footnote stands for the relative dielectric permittivity and relative magnetic permeability respectively. The relative dielectric permittivity of air is close to 1. That of sand particles can be obtained from literature value. The relative magnetic permeability for both air and sand are around 1. By matching the measured data with that predicted by Equation, the term $\varepsilon_{r,FeO}\mu_{r,FeO}$ is estimated to be around 116.1. As the dielectric constant of FeO is approximately constant at microwave frequency (around 16.6), the relative magnetic permeability is estimated to be around 7 at equivalent TDR frequency (low gigahertz range).

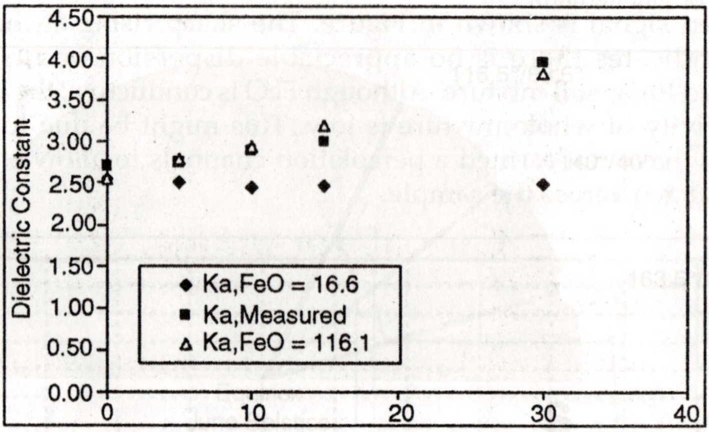

Fig. FeO Content by Weight (%)

The following conclusions were drawn based on this investigation: A detailed analyses indicated that the magnetic permeability for FeO at TDR frequency range is around. The following propose a hypothesis to describe the origination of cohesion in lunar soil. The Moon, similar as the Earth, is a huge magnet. Ferromagnetic material Fe-particle in lunar soils tends to be magnetized under the magnetic field of Moon.

They act as small magnets, which tend to attract with each other. The closer the distance, the larger the attraction forces. A method to differentiate their contribution to cohesion from other sources such as molecular attractions is if mixing them randomly, the cohesion will disappear due to disturbance of the alignments of these "small magnets". The cohesion will increase, however, as Fe grains are re-magnetized at their new locations.

Chapter 6

Global Tectonics

Plate Tectonics is a theory developed in the late 1960s, to explain how the outer layers of the Earth move and deform. The theory has caused a revolution in the way we think about the Earth. Since the development of the theory, geologists have had to reexamine almost every aspect of Geology. Plate tectonics has proven to be so useful that it can predict geologic events and explain almost all aspects of what we see on the Earth. Although we have discussed plate tectonics throughout the course, in this lecture we look at how the theory came to be discovered and some of its implications for the evolution of the Earth.

TECTONIC THEORIES

Tectonic theories attempt to explain why mountains, earthquakes, and volcanoes occur where they do, the ages of deformational events, and the ages and shapes of continents and ocean basins.

- Late 19th Century Theories
  - Contraction of the Earth due to cooling. This is analogous to what happens to the skin of an apple as the interior shrinks as it dehydrates. It could explain compressional features, like fold/thrust mountain belts, but could not explain extensional features, such as rift valleys and ocean basins. Nor could it explain the shapes and positions of the continents.
  - Expansion of the Earth due to heating. This was suggested after radioactivity was discovered. This could explain why the continents are broken up, and could easily explain extensional features, but did not do well at explaining compressional features.
- Wegner's Theory of Continental Drift

Alfred Wegner was a German Meteorologist in the early 1900s who studied ancient climates. Like most people, the jigsaw puzzle appearance of the Atlantic continental margins caught his attention.

He put together the evidence of ancient glaciations and the distribution of fossil to formulate a theory that the continents have moved over the surface of the Earth, sometimes forming large supercontinents and other times forming separate continental masses. He proposed that prior to about 200 million years ago all of the continents formed one large land mass that he called Pangea.

The weakness of Wegner's theory, and the reason it was not readily accepted by geologists was that he proposed that the continents slide over ocean floor. Geophysicists disagreed, stating the ocean floor did not have enough strength to hold the continents and too much frictional resistance would be encountered.

In 1950s and 1960s, studies of the Earth's magnetic field and how it varied through time *(paleomagnetism)* provided new evidence that would prove that the continents do indeed drift. In order to understand these developments, we must first discuss the Earth's magnetic field and the study of Paleomagnetism.

PALEOMAGNETISM

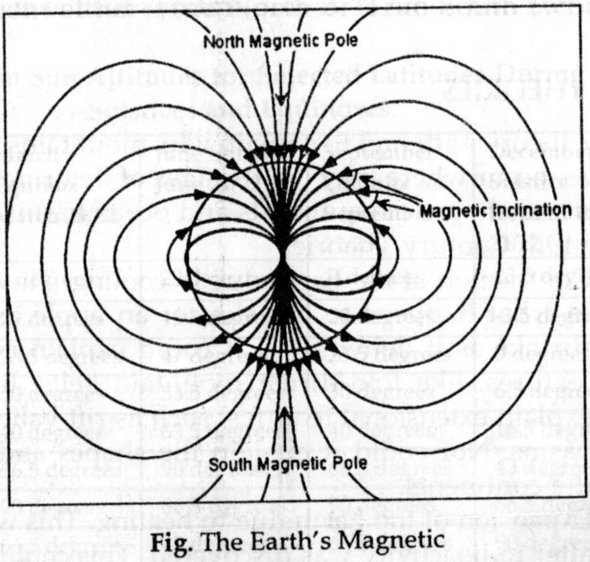

Fig. The Earth's Magnetic

The Earth has a magnetic field that causes a compass needle to always point toward the North magnetic pole, currently located near the rotation pole. The Earth's magnetic field is what would be expected if there were a large bar magnet located at the centre of the Earth (we now know that this is not what causes the magnetic field, but the analogy is still good).

A compass needle or a magnetic weight suspended from a

string, points along these lines of force. Note that the lines of force intersect the surface of the Earth at various angles that depend on position on the Earth's surface. This angle is called the *magnetic inclination*. The inclination is 0° at the magnetic equator and 90° at the magnetic poles. Thus, by measuring the inclination and the angle to the magnetic pole, one can tell position on the Earth relative to the magnetic poles.

In the 1950s it was discovered that when magnetic minerals cool below a temperature called the *Curie Temperature*, domains within the magnetic mineral take on an orientation parallel to any external magnetic field present at the time they cooled below this temperature.

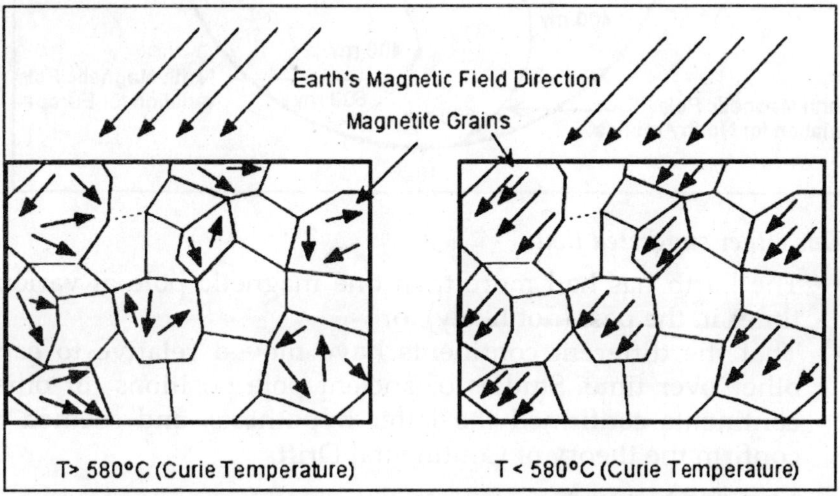

At temperatures above the Curie Temperature, permanent magnetization of materials is not possible. Since the magnetic minerals take on the orientation of the magnetic field present during cooling, we can determine the orientation of the magnetic field present at the time the rock containing the mineral cooled below the Curie Temperature, and thus, be able to determine the position of the magnetic pole at that time. This made possible the study of Paleomagnetism (the history of the Earth's magnetic field). Magnetite is the most common magnetic mineral in the Earth's crust and has a Curie Temperature of 580°C

Initial studies of the how the position of the Earth's magnetic pole varied with time were conducted in Europe. These studies showed that the magnetic pole had apparently moved through time. When similar measurements were made on rocks of various ages in North America, however, a different path of the magnetic pole was found.

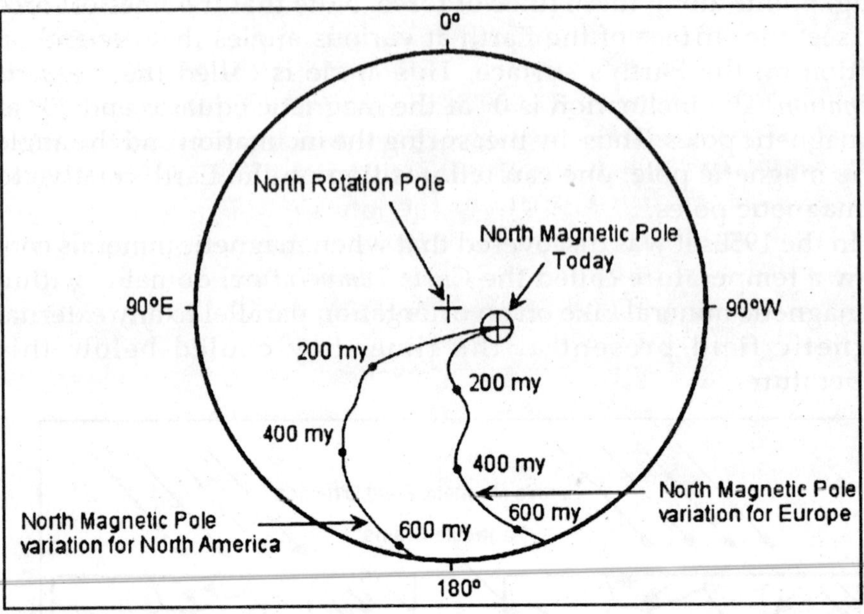

This either suggested that:

- The Earth has had more than one magnetic pole at various times in the past (not likely), or
- That the different continents have moved relative to each other over time. Studies of ancient pole positions for other continents confirmed the latter hypothesis, and seemed to confirm the theory of Continental Drift.

SEA-FLOOR SPREADING

During World War II, geologists employed by the military carried out studies of the sea floor, a part of the Earth that had received little scientific study. The purpose of these studies was to understand the topography of the sea floor to find hiding places for both Allied and enemy submarines. The topographic studies involved measuring the depth to the sea floor. These studies revealed the presence of two important topographic features of the ocean floor:

- *Oceanic Ridges:-* Long sinuous ridges that occupy the middle of the Atlantic Ocean and the eastern part of the Pacific Ocean.
- *Oceanic Trenches*: deep trenches along the margins of continents, particularly surrounding the Pacific Ocean.

Another type of study involved towing a magnetometer (for measuring magnetic materials) behind ships to detect submarines.

The records from the magnetometers, however, revealed that there were magnetic anomalies on the sea floor, with magnetic high areas running along the oceanic ridges, and parallel bands of alternating high and low magnetism on either side of the oceanic ridges. Before these features can be understood, we need to first discuss another development in the field of Paleomagnetism - the discovery of reversals of the Earth's magnetic field and the magnetic time scale.

- Reversals of the Earth's Magnetic Field. Studying piles of lava flows on the continents geophysicists found that over short time scales the Earth's magnetic field undergoes polarity reversals (The north magnetic pole becomes the south magnetic pole) By dating the rocks using radiometric dating techniques and correlating the reversals throughout the world they were able to establish the magnetic time scale.

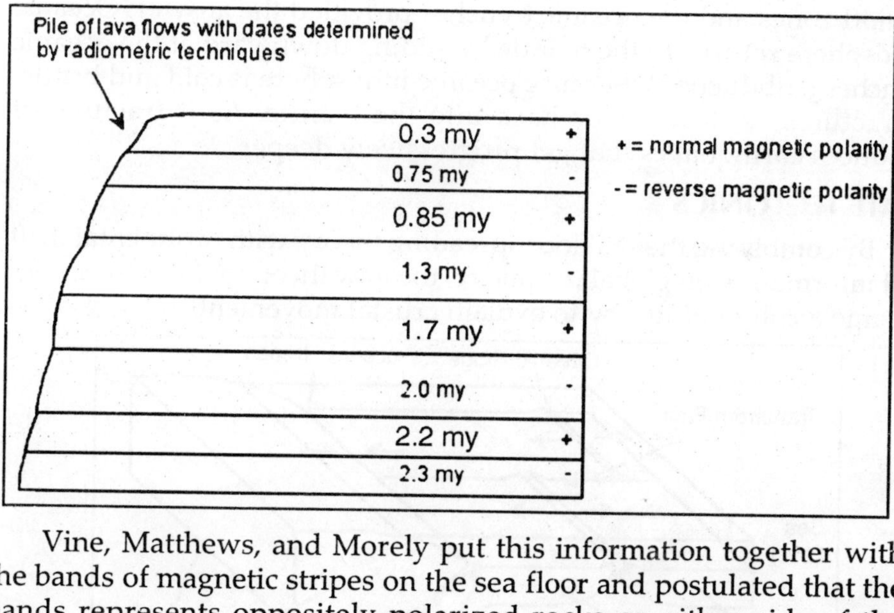

Vine, Matthews, and Morely put this information together with the bands of magnetic stripes on the sea floor and postulated that the bands represents oppositely polarized rocks on either side of the oceanic ridges, and that new oceanic crust and lithosphere was created at the oceanic ridge by eruption and intrusion of magma. As this magma cooled it took on the magnetism of the magnetic field at the time.

When the polarity of the field changed new crust and lithosphere created at the ridge would take on the different polarity. This hypothesis led to the theory of sea floor spreading.

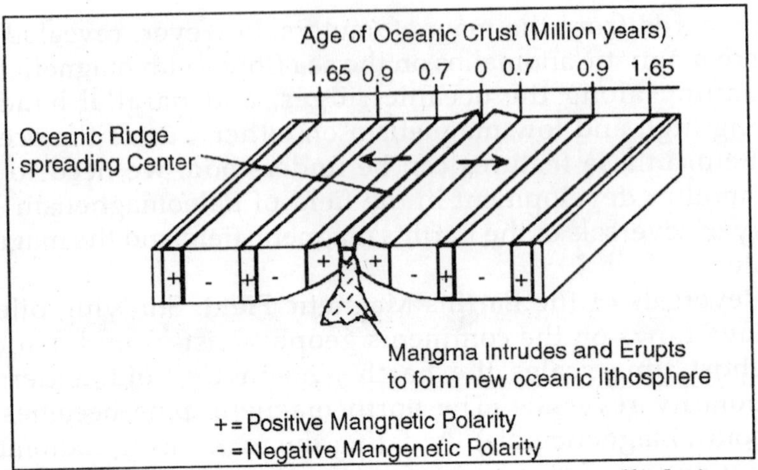

If new oceanic crust and lithosphere is continually being created at the oceanic ridges, the oceans should be expanding indefinitely, unless there were a mechanism to destroy the oceanic lithosphere. Benioff zones and the oceanic trenches provided the answer: Oceanic lithosphere returns to the mantle by sliding downward at the oceanic trenches (subducting). Because oceanic lithosphere is cold and brittle, it fractures as it descends back into the mantle. As it fractures it produces earthquakes that get progressively deeper.

PLATE TECTONICS

By combining the sea floor spreading theory with continental drift and information on global seismicity, the new theory of Plate Tectonics became a coherent theory to explain crustal movements.

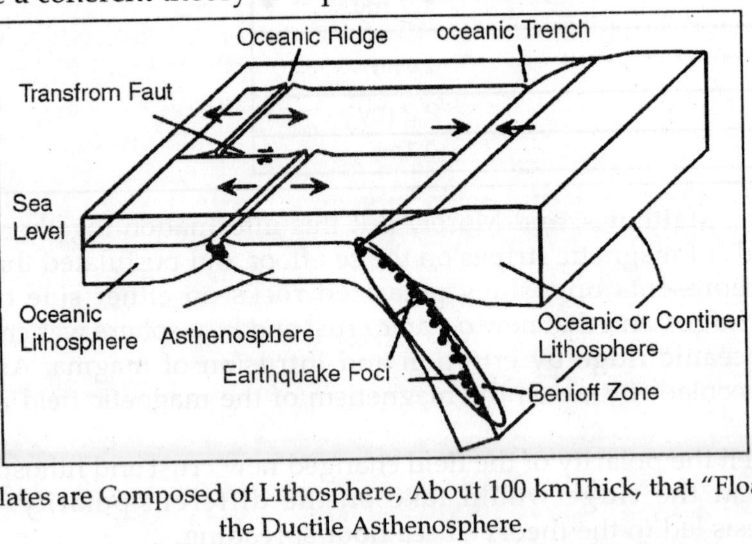

Fig. Plates are Composed of Lithosphere, About 100 kmThick, that "Float" on the Ductile Asthenosphere.

While the continents do indeed appear to drift, they do so only because they are part of larger plates that float and move horizontally on the upper mantle asthenosphere. The plates behave as rigid bodies with some ability to flex, but deformation occurs mainly along the boundaries between plates.

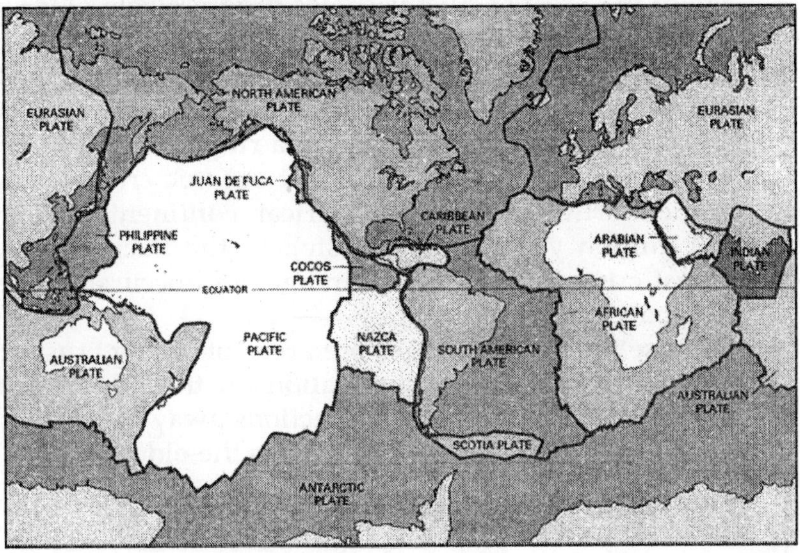

TYPES OF PLATE BOUNDARIES

Divergent Plate Boundaries

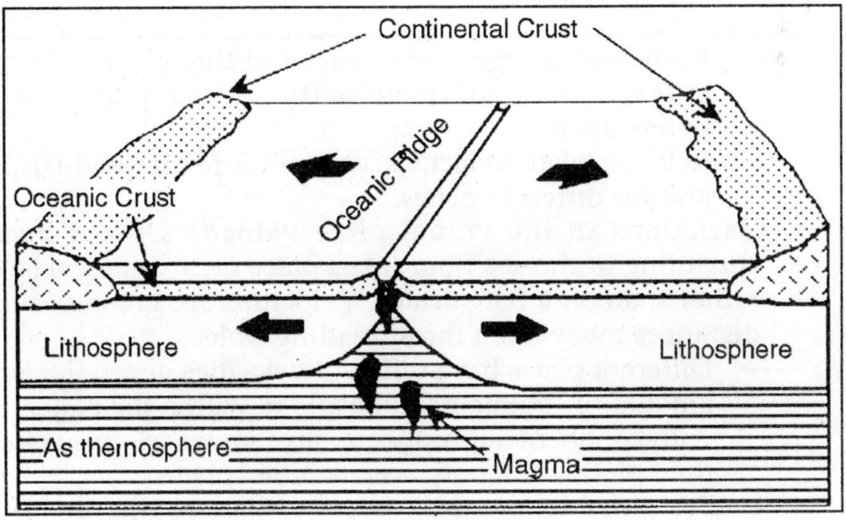

Fig. Diverging Plate Boundary Oceanic Ridge - Spreading Cevter

- These are oceanic ridges where new oceanic lithosphere is created by upwelling mantle that melts, resulting in basaltic magmas which intrude and erupt at the oceanic ridge to create new oceanic lithosphere and crust. As new oceanic lithosphere is created, it is pushed aside in opposite directions. Thus, the age of the oceanic crust becomes progressively older in both directions away from the ridge.
- Because oceanic lithosphere may get subducted, the age of the ocean basins is relatively young. The oldest oceanic crust occurs farthest away from a ridge.
  In the Atlantic Ocean, the oldest oceanic crust occurs next to the North American and African continents and is about 180 million years old (Jurassic). In the Pacific Ocean, the oldest crust is also Jurassic in age, and occurs off the coast of Japan.
- Because the oceanic ridges are areas of young crust, there is very little sediment accumulation on the ridges. Sediment thickness increases in both directions away of the ridge, and is thickest where the oceanic crust is the oldest.

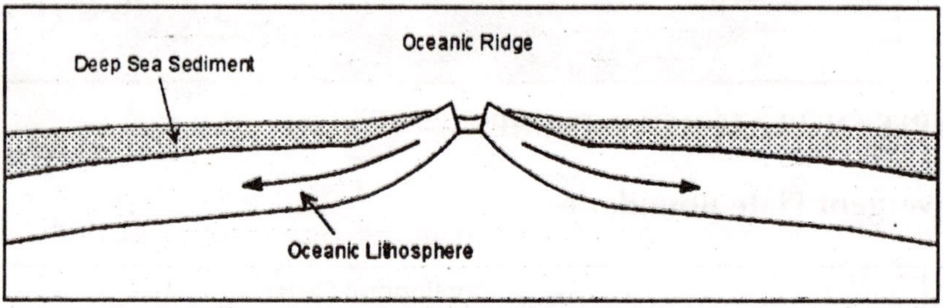

  - Knowing the age of the crust and the distance from the ridge, the relative velocity of the plates can be determined.
  - Relative plate velocities vary both for individual plates and for different plates.
- Variations in individual plate velocities occur because spreading of the sea floor takes place on a spherical surface rather than on a flat surface. Velocities are greatest at large distances away from the spreading pole.
  - Different plates have different velocities depending on the amount of continental lithosphere within the plate. Plates with continental lithosphere have lower relative velocities than plates with only oceanic lithosphere.
  - Sea floor topography is controlled by the age of the oceanic lithosphere and the rate of spreading.

If the spreading rate (relative velocity) is high, magma must be rising rapidly and the lithosphere is relatively hot beneath the ridge. Thus for fast spreading centers the ridge stands at higher elevations than for slow spreading centers. The rift valley at fast spreading centers is narrower than at slow spreading centers.

As oceanic lithosphere moves away from the ridge, it cools and sinks deeper into the asthenosphere. Thus, the depth to the sea floor increases with increasing age away from the ridge.

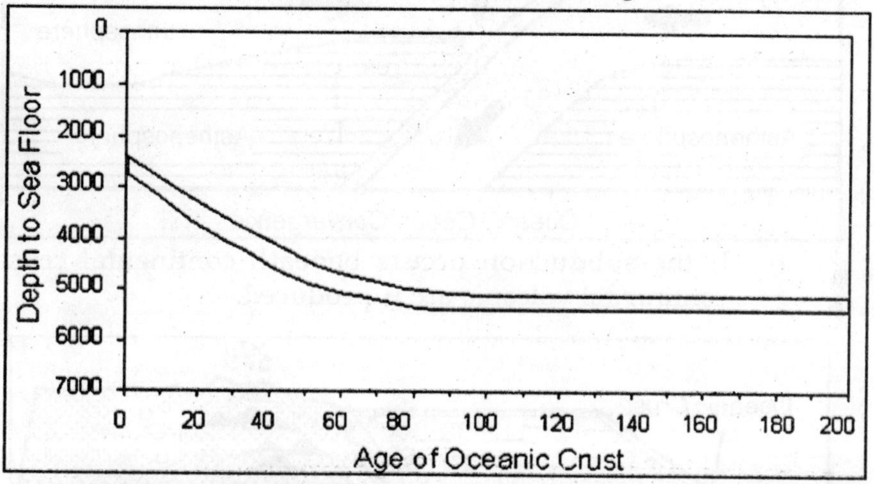

Convergent Plate Boundaries

- When a plate of dense oceanic lithosphere moving in one direction collides with a plate moving in the opposite direction, one of the plates subducts beneath the other. Where this occurs an oceanic trench forms on the sea floor and the sinking plate becomes a subduction zone. The Benioff zone identifies a subduction zone. The earthquakes may extend down to depths of 700 km before the subducting plate heats up and loses its ability to deform in a brittle fashion.
- As the oceanic plate subducts, it begins to heat up and metamorphose.

  As it does so, dehydration reactions release water into the overlying mantle asthenosphere, causing a reduction in the melting temperature and the production of andesitic magmas. These magmas rise to the surface and create a volcanic arc parallel to the trench.
  - If the subduction occurs beneath oceanic lithosphere, an island arc is produced at the surface.

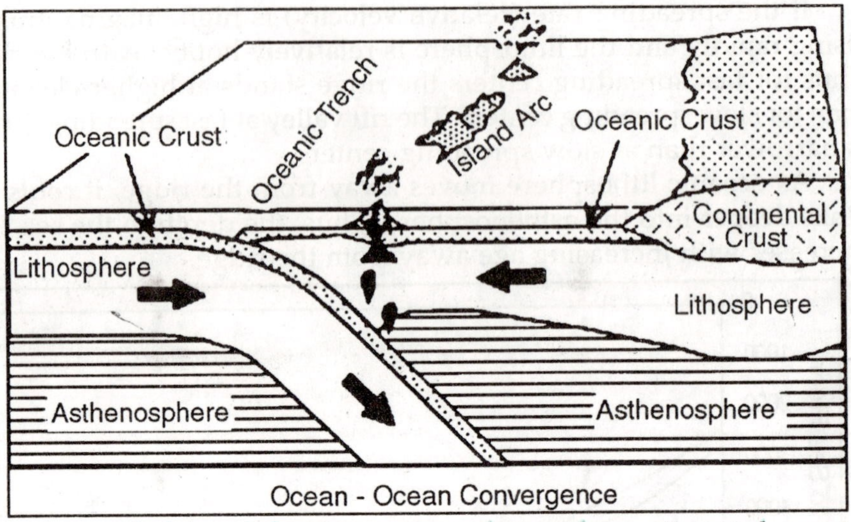

- If the subduction occurs beneath continental crust, a continental volcanic arc is produced.

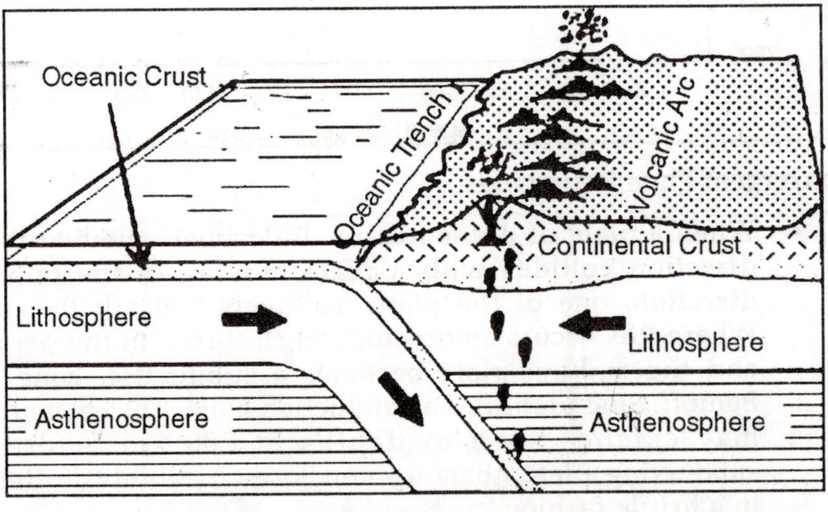

Fig. Ocean - Continent Convergence

- If one of the plates has continental lithosphere on its margin, the oceanic plate will subduct because oceanic lithosphere has a higher density than continental lithosphere.
  - Sediment deposited along the convergent margin, and particularly that in the trench will be deformed by thrust faulting. This will break the rocks up into a chaotic mixture of broken, jumbled, and thrust faulted rock know as a *mélange*.

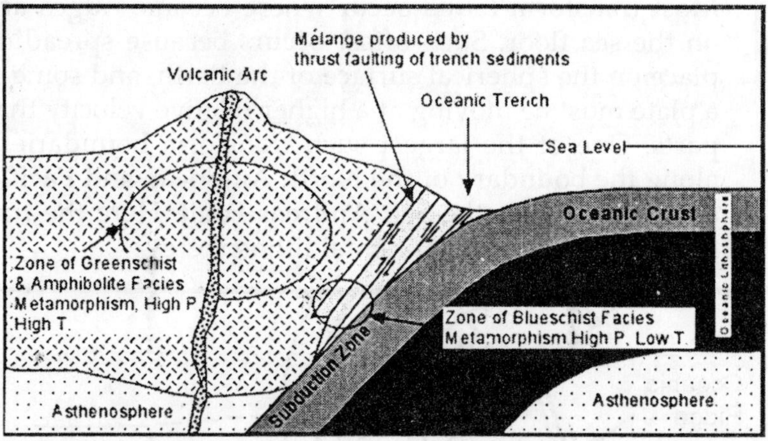

- Since the mélange is relatively cool, as it is pushed deeper, the pressure increases and it becomes metamorphosed under conditions of high pressure and low temperature, resulting in blueschist facies metamorphism. Beneath the magmatic arc, the passage of magma through the overlying plate results in increased temperatures and metamorphism at high pressures and high temperatures producing greenschist and amphibolite facies metamorphic rocks.

Transform Plate Boundaries

Where lithospheric plates slide past one another in a horizontal manner, a transform fault is created. Earthquakes along such transform faults are shallow focus earthquakes.

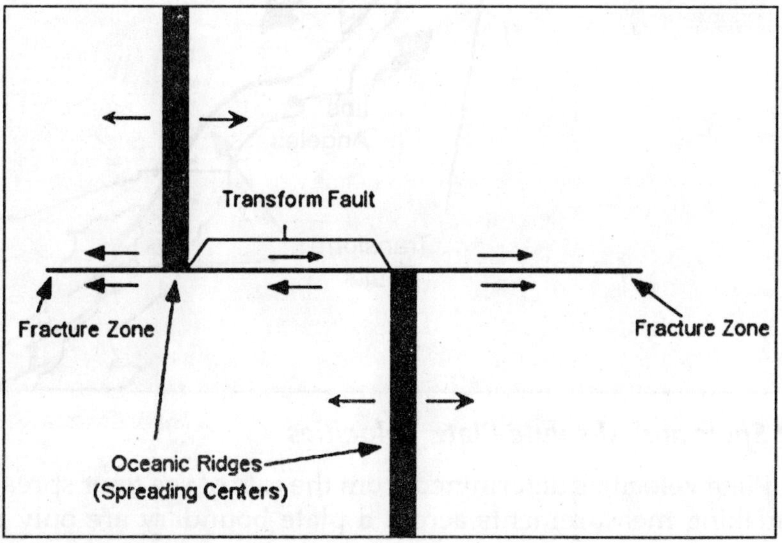

- Most transform faults occur where oceanic ridges are offset on the sea floor. Such offset occurs because spreading takes place on the spherical surface of the Earth, and some parts of a plate must be moving at a higher relative velocity than other parts. One of the largest such transform boundaries occurs along the boundary of the North American and Pacific plates and is known as the San Andreas Fault. Here the transform fault cuts through continental lithosphere

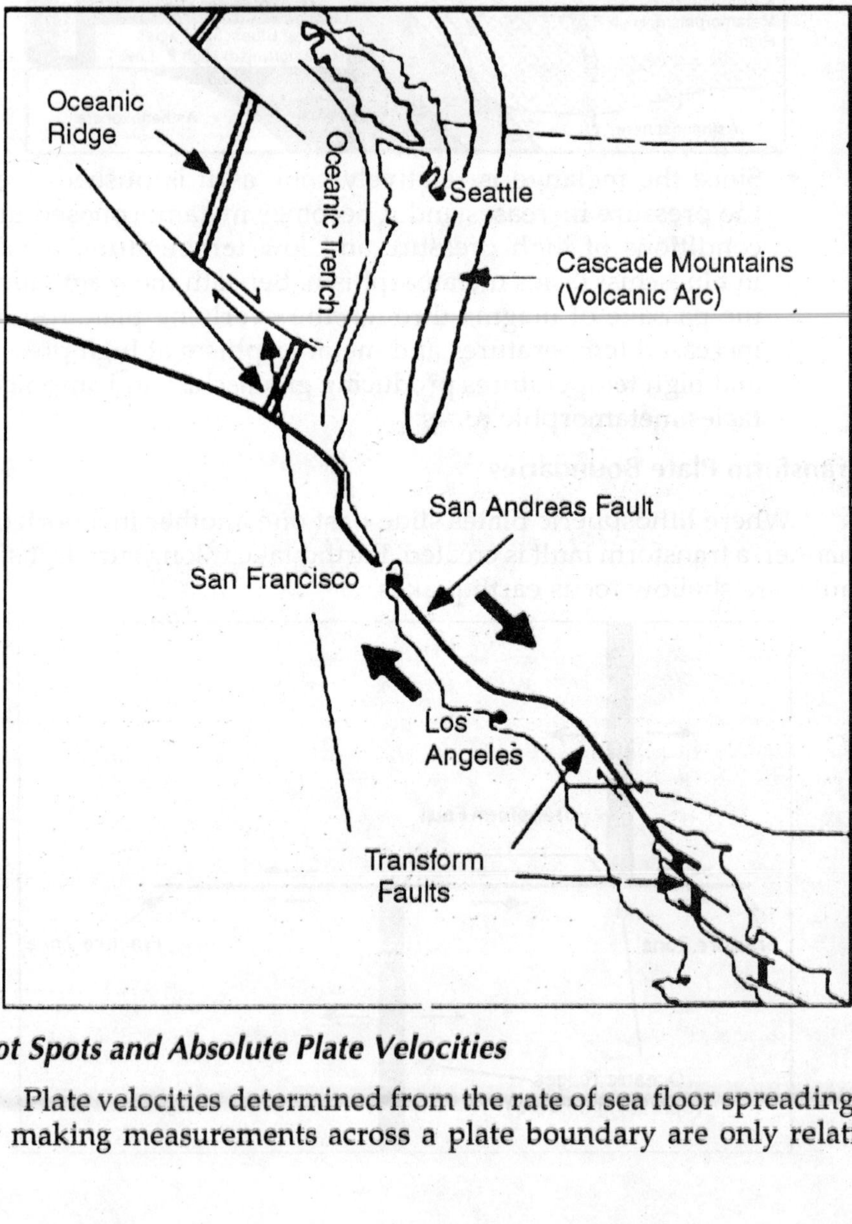

Hot Spots and Absolute Plate Velocities

Plate velocities determined from the rate of sea floor spreading or by making measurements across a plate boundary are only relative velocities. That is we know the velocity of one plate only if we can assume that the adjacent plate is not moving. In order to determine absolute plate velocities, we need some fixed reference point that we know is not moving.

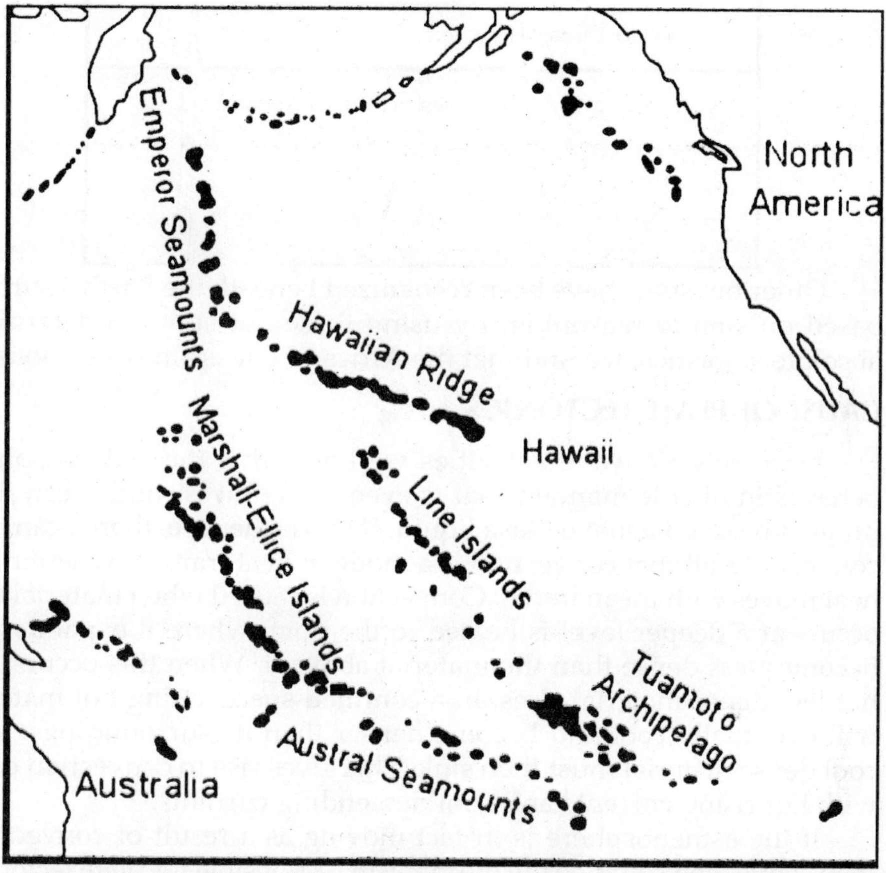

One place where this might be possible is in the Pacific Ocean, where the Hawaiian Islands are part of a chain of islands, far removed from any plate boundary, where islands and seamounts in the chain increase in age from the southeast to the northwest. Furthermore, the island at the southeast end of the chain, the Big Island of Hawaii, is the only island with currently active volcanoes. The island chain appears to have formed as the Pacific plate moved over a *Hot Spot*, an area in the Earth's mantle where hot material from the Earth's interior is moving upward.

If we can assume that such a hot spot is stationary, then we can calculate the absolute velocity of the Pacific Plate as it has moved over the hot spot.

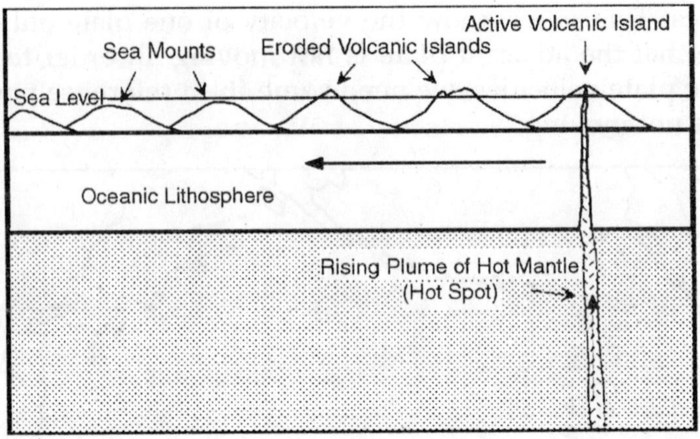

Other hot spots have been recognized beneath the Earth's surface based on similar reasoning. By using these hot spots to determine absolute velocities, we find that the African Plate is almost stationary.

CAUSE OF PLATE TECTONICS

From seismic wave velocities we know that the asthenosphere behaves in ductile manner, that is even though it is solid it can flow under stress and behave like a liquid. If this is the case, then it can also convect. Recall that convection is a mode of heat transfer wherein the heat moves with the material. Convection is caused when material that occurs at a deeper level is heated to the point where it expands and becomes less dense than the material above it. When this occurs, the hot less dense material rises. In a confined space, rising hot material will eventually cool and become denser than its surroundings. This cool dense material must then sink. This gives rise to convection cells, with hot rising currents and cool descending currents.

If the asthenosphere is in fact moving as a result of convection, then convection could be the mechanism responsible for plate tectonics. Hot rising currents would occur beneath oceanic ridges.

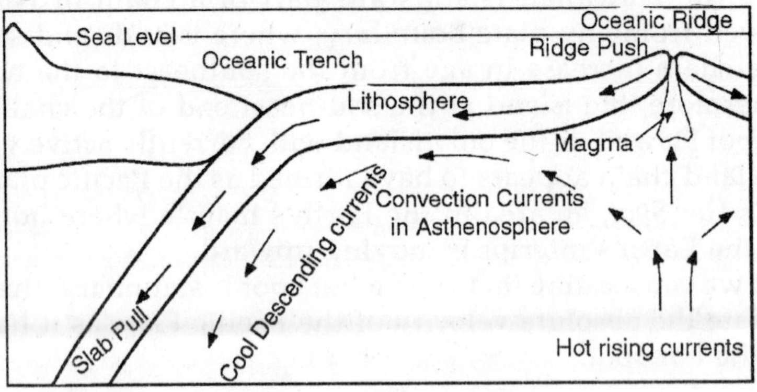

Magma intruding into the ridge would push lithosphere apart at the ridge. As the new lithosphere cools, it will slide off the topographic high that results from the upwelling of the mantle and will eventually become cold and dense. This dense lithosphere will tend to pull the rest of the lithosphere downward. A combination of dragging the lithosphere along the top of the convection cell, ridge push, sliding, and slab pull all appear to be contributing factors to the cause of plate tectonics.

Plate Tectonics and Continental Crust

The continents can be divided into two kinds of structural units

- *Cratons:* Form the cores of the continents. These are portions of continental crust that have attained isostatic and tectonic stability. They were formed and were deformed very early in Earth history and are the oldest parts of the continents.
- *Orogens:* Are broad elongated belts of deformed rocks that are draped around the cratons. They appear to be the eroded roots of former mountain belts that formed by continent - continent collisions. Only the youngest of these orogens still form mountain ranges.

The observation that the orogens are generally younger towards the outside of any continent suggests that the continents were built by collisions of plates that added younger material to the outside edges of the continents, and is further evidence that plate tectonics has operated for at least the last 2 billion years. Because continents can join by collision and can split by rifting, we need to examine the various types of continental margins that can occur and see how they relate to plate boundaries.

Passive Continental Margins

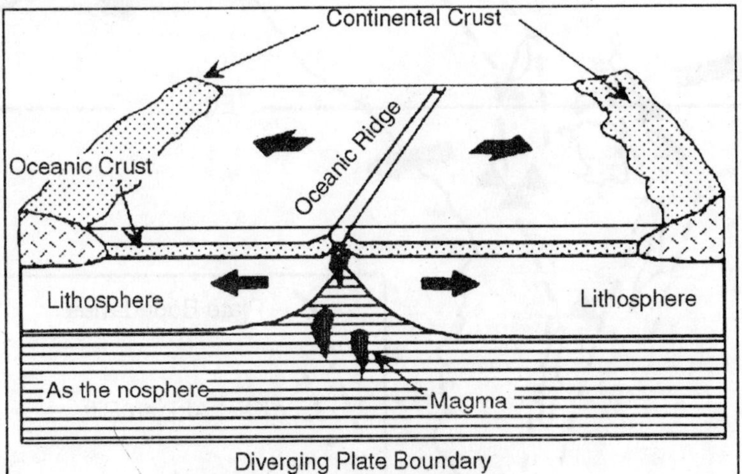

Diverging Plate Boundary
Oceanic Ridge - Spreading Center

A passive continental margin occurs in the interior of plate, far away from any plate boundary. Present examples of passive continental margins are the Atlantic coast of North and South America, Europe, and Africa. No current deformation is taking place along these margins because they are not close to plate boundaries. The passive continental margins developed as a result of rifting of a former larger continent.

Areas where rifting may be forming future passive continental margins include the Red Sea area of northern Africa, the East African Rift Valley, and the Basin and Range Province of the Western U.S. Block fault mountains are common in the initial stages of rifting.

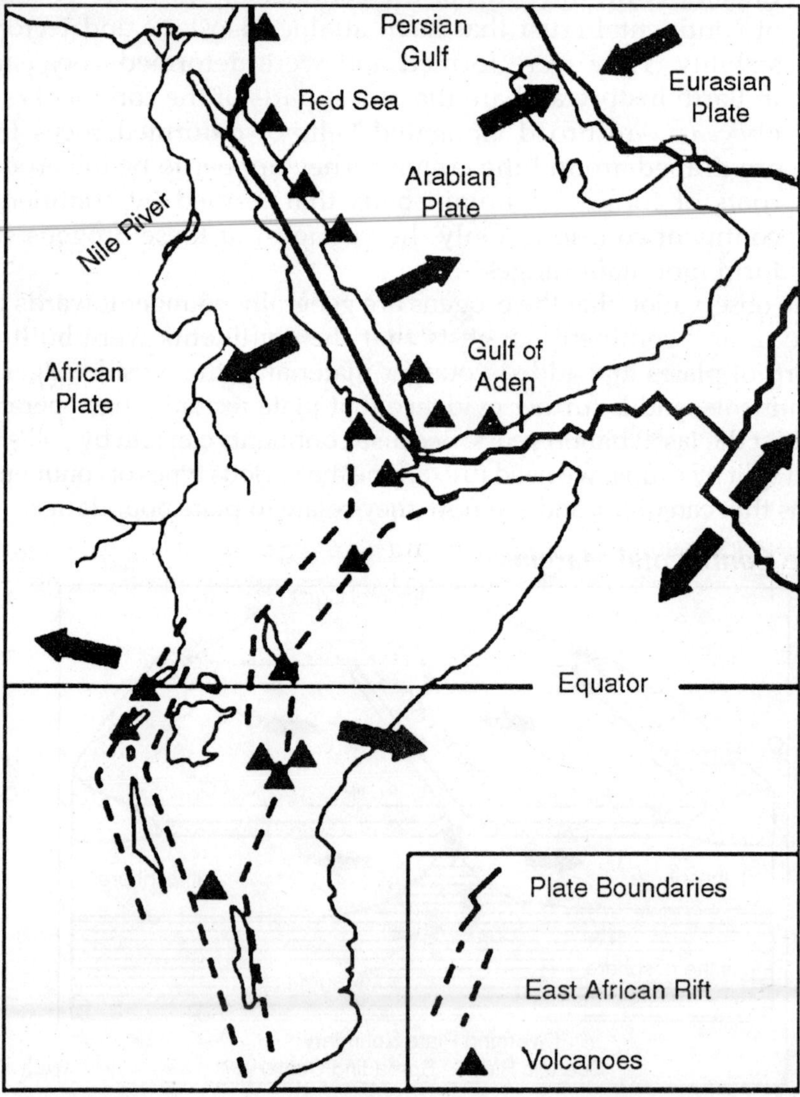

- *Continental Convergent Margins*: Continental convergent margins occur where the margin of the continent coincides with a convergent plate boundary.

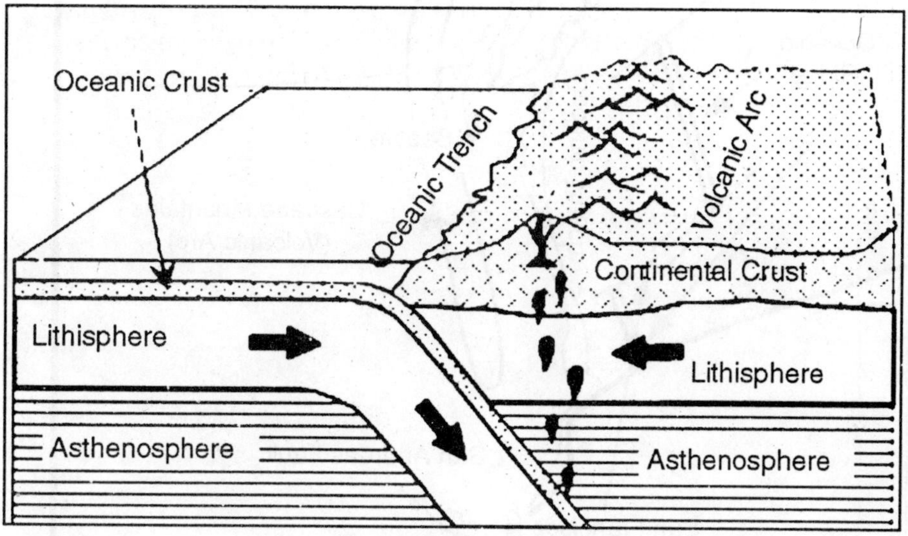

Fig. Ocean - Continent Convergence

Subduction of the oceanic lithosphere beneath the continental lithosphere produces continental volcanic arcs that erupt mostly andesitic magma.

Sediments along the margin are deformed into mélanges and a pair of metamorphic belts develops beneath the continental margin, the one closest to the plate margin shows blueschist facies metamorphism and the one beneath the volcanic arc shows greenschist and amphibolite facies metamorphism.

When such paired metamorphic belts are observed in ancient rocks, this provides evidence that the area was once a convergent continental margin.

The best example of a current convergent continental margin occurs along the Pacific coast of South America and in the Cascade Mountains of the western U.S.

Transform Fault Margins

When a continental margin coincides with a transform fault plate boundary the continental margin is called a transform fault margin. As discussed before, the boundary between the Pacific Plate and the North American Plate in the western U.S. is such a plate boundary formed by the San Andreas Fault. The San Andreas Fault developed beginning in the Mesozoic when a ridge separating the two plates was subducted.

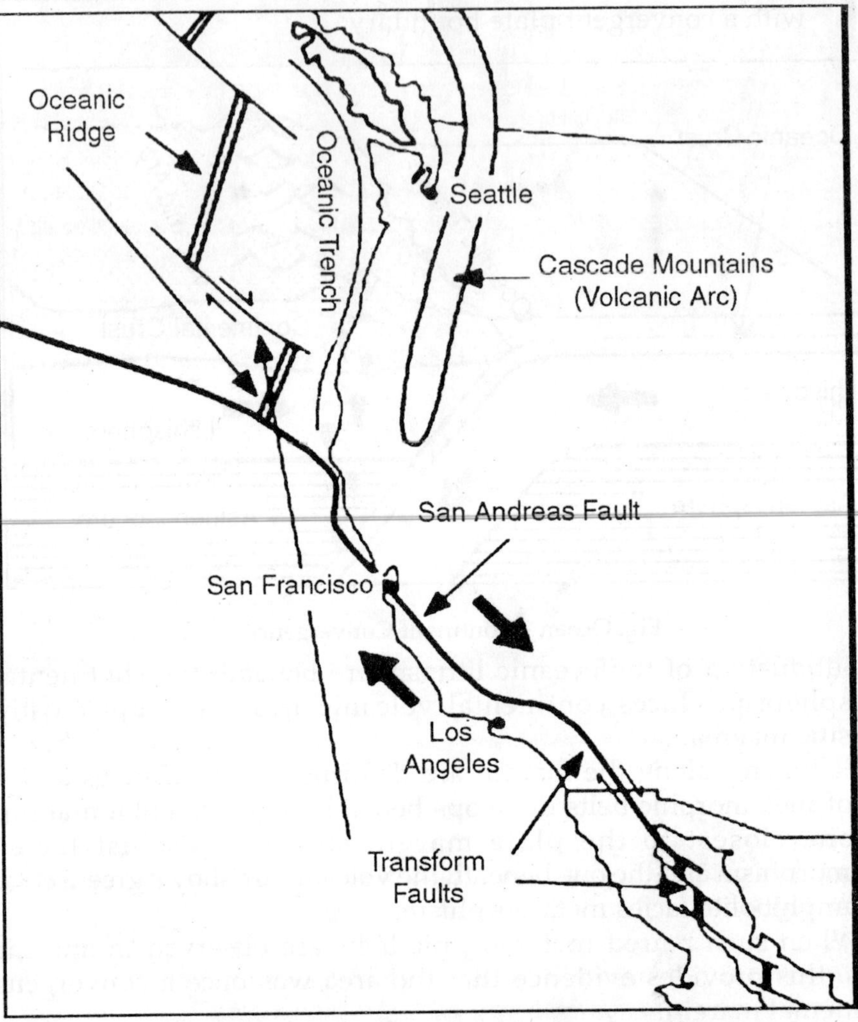

Continental Collision Margins

When two plates that have continental lithosphere collide with one another the margin becomes a continental collision margin. Such margins are characterized by fold - thrust mountain belts that develop along the zone of collision. Currently the highest mountains in the world, the Himalayas represent this kind of margin.

The Himalayas resulted from a collision of the plate containing India with the plate containing Eurasia. This collision is still taking place and results in joining the two formerly separate plates. The occurrence of ancient fold -thrust mountain belts such as the Appalachian Mountains of the Eastern U.S., the Urals of Central Russia, and the Alps of southern Europe, are evidence of ancient continental collision margins.

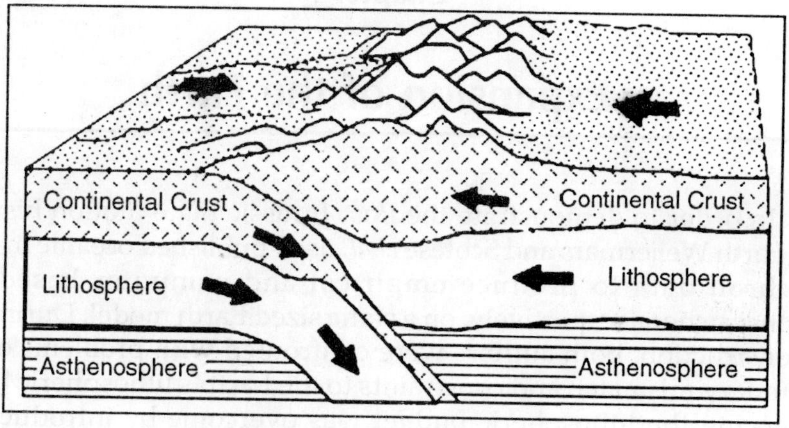

Fig. Continent - Continent Convergence

Accreted Terrane Margins

A former convergent margin or transform fault margin that has been modified by numerous additions of small blocks of crust that have accreted to the continental margin is called an accreted terrane margin. A terrane is a package of rocks with a distinctive stratigraphic and structural history that appears to have formed at some location other than its present location. The northwestern margin of North America is composed of many such terranes that appear to have been added to the continent. An accreted terrane margin occurs close to a plate boundary because it once was a plate boundary of the convergent or transform type.

Chapter 7

Expansion of the Earth

In order to demonstrate the Post-Jurassic plate motion history of the Earth Weijermars and Scotese *et al,* used published oceanic magnetic isochron data to produce empirical and computer based plate reconstructions respectively, on a static sized Earth model. During plate reconstruction, both authors were confronted with problems of plate boundary mismatch and constraints to the Earth's lithospheric budget. Balancing the lithospheric budget was overcome by introduction of presumed, pre-existing lithosphere in order to maintain a static Earth radius. In order to justify modeling on a static Earth radius, Weijermars in particular, provided "six conclusive arguments against any fast Earth expansion hypothesis", prior to modeling studies. This, unfortunately, did not scientifically resolve the argument that modern oceanic magnetic isochron data may in fact be better modeled at reduced Earth radii.

Estimates of palaeoradius from palaeomagnetic data were, and still are, considered by palaeomagneticians as limiting the amount of potential Earth expansion to less than 0.8% during the last 400 million years. Crater distributions and ages on the lunar surface were also considered to rule out any significant expansion of the Earth or Moon during the last 4 billion years, and similar conclusions were applied to Mars, Mercury and Venus. Similarly biogeographic distributions of invertebrates during the early Palaeozoic were considered to be incompatible with an expanding Earth. Limits to palaeogravity since the late Precambrian were said to contradict rapid Earth expansion, and annual growth increments of corals suggested to van Diggelen that the Earth has not expanded.

Direct measuring across the Pacific Ocean, to determine the relative plate motions, began in 1976 when NASA launched the Laser Geodynamics Satellite (LAGEOS), as part of their Geodynamics Programme, Crustal Dynamics Project. The LAGEOS laser ranging data obtained up to 1984, combined with the LAGEOS data derived from Christodoulidis *et al,* plus more recent VLBI and SLR geodetic measurements indicated convergence rates in the Pacific significantly less than those predicted by Minster & Jordan, based on a theoretical mathematical model on a constant sized Earth. In particular the chord length increase between Australia and South America. Owen and Carey indicated that the accountancy of these preliminary data do not balance as they should if the Earth were of constant dimensions.

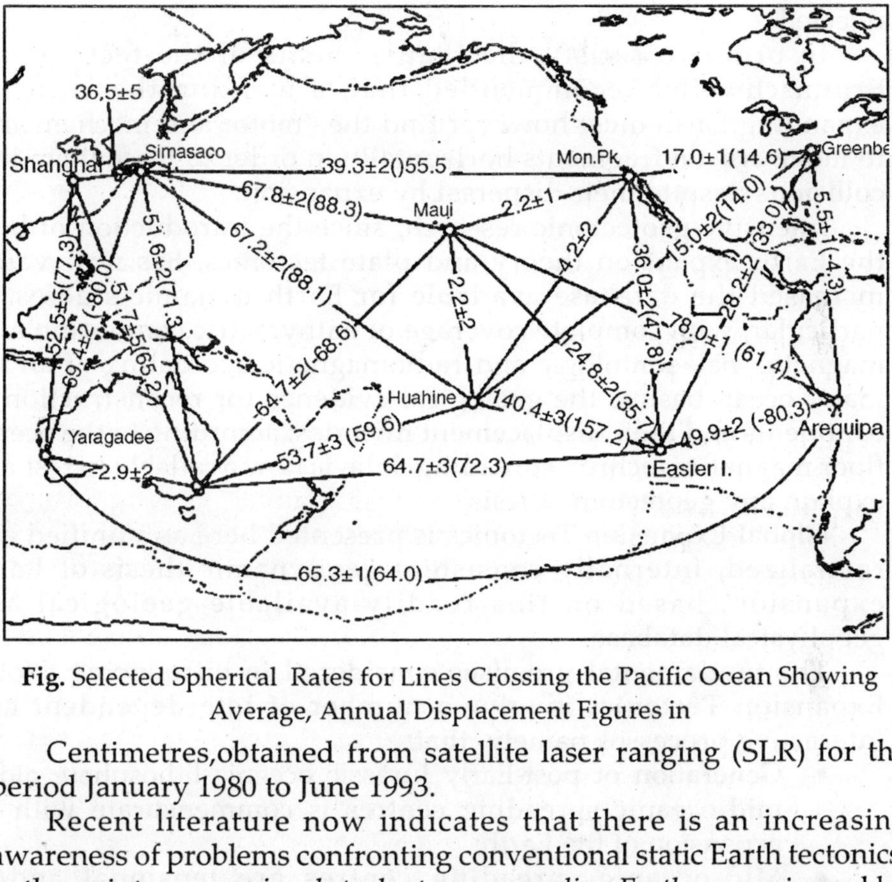

Fig. Selected Spherical Rates for Lines Crossing the Pacific Ocean Showing Average, Annual Displacement Figures in

Centimetres,obtained from satellite laser ranging (SLR) for the period January 1980 to June 1993.

Recent literature now indicates that there is an increasing awareness of problems confronting conventional static Earth tectonics, to the point were *major obstacles to an expanding Earth,* as envisaged by McElhinny *et al* and Schmidt & Clark, do not necessarily *"outnumber the evidence in favour"*. Of particular interest are the findings of Ahmad who demonstrated that palaeontological and palaeomagnetic data show unequivocally that India was not involved in long distance drift and collision with the northern continents, as advocated by plate tectonics. Ahmad suggested instead that Pangaea existed as a large, composite super-continent comprising the entire existing continental crust. Palaeomagnetic pole positions indicated to Ahmad that, during the Upper Permian the Earth was 55% to 60% of its present diameter.

Similarly, at the Smithsonian Institution discussion meeting in July 1989, numerous arguments were presented concerning new concepts in global tectonics.

These arguments indicated that there seems to be something questionable with the plate tectonic theory as it is currently presented, and that present concepts of plate tectonics - continental drift - polar wandering may need to be re-evaluated, revised, or rejected.

In order to satisfy the requirements of the tectonician, Brunnschweiler recommended that, a working thesis for an expanding Earth must however find the "motor and mechanisms" to move crustal fragments horizontally in order to force them into collisions despite their dispersal by expansion.

Intensive geotectonic research, since the introduction of both the Earth expansion theory and plate tectonics, has now vastly increased the database available for Earth dynamic studies. In particular, with complete coverage of bathymetric contouring and, magnetic, palaeontologic and radiomagnetic age dating of all the major ocean basins, the geological evidence for reconstruction of continents and their displacement histories, according to the ocean-floor magnetic isochron spreading data, is now available to test and explain any geotectonic thesis.

Global Expansion Tectonics is presented here as a unified and revitalized, internally consistent geodynamic thesis of Earth expansion, based on this readily available geological and geophysical database.

The fundamental premises considered in introducing Global Expansion Tectonics involve a number of interdependent and interactive processes namely, that:

- Generation of post-Early Jurassic oceanic lithosphere along mid-oceanic spreading centres is commensurate with an expansion of the Earth;
- Mid-oceanic spreading centres are tensional and/or transtensional règimes, with oceanic lithosphere generated in response to a separation of plate margins in sympathy with an expansion of the Earth;
- New oceanic lithosphere, generated along mid-oceanic spreading centres, is cumulative with time, with no requirement for removal of older lithosphere by subduction processes;
- The total surface area of oceanic lithosphere accumulated in a given time interval is a reflection of the increase in

palaeoradius during that time interval, and is therefore a measurable quantity;

- Basin development and orogenesis are natural consequences of an expanding Earth, as continental and oceanic lithosphere re-equilibrate to a changing surface curvature, and;
- Hydrospheric and atmospheric accumulation has been increasing with time, in sympathy with an accumulation of the lithosphere.

EMPIRICAL SMALL EARTH MODELING

The Earth, like any sphere, does not lend itself to an undistorted representation of its surface area upon a flat map or page of a book. Maps however, are intrinsically necessary for the presentation of Earth based information.

Because of this, various mathematical equations have been devised over the years in order to project the image of the surface features of the Earth into map format. The distortions, inherent in all of these cartographic projection techniques are, out of necessity, generally accepted, and acknowledged in the presentation of map based information.

In the presentation of expanding Earth tectonic processes however, as the diameter of the small Earth globe is reduced backwards in time, the distortion of continental outlines becomes increasingly unrealistic and unconvincing in map format as continental areas progressively occupy more of the surface area of the globe.

To represent the entire Earth it is therefore necessary to rely heavily on spherical small Earth global models to define accurately the continental configurations, and plate motion history, throughout geological time.

This historical development of small Earth global modeling is briefly summarised from Carey, and Vogel, in order to familiarise the reader with the work of past model makers.

The historical development of the theory of Earth expansion has been comprehensively covered by Carey and will not be addressed further.

Hilgenberg: Stimulated by the pioneering work of Alfred Wegener on continental drift, Otto Hilgenberg has been attributed as being the first to fit all the land masses together to completely enclose a small papier-machè globe. On his globes all oceans were eliminated and the sialic crust neatly enclosed the whole Earth on a globe about 60% of the diameter of the present Earth.

Fig. Hilgenberg's Expanding Earth "Terrella" Attributed to Being the First Small Earth Models Constructed. Small globe is Approximately 60% of Present Earth Radius.

Reconstruction across the Atlantic was considered to be convincing however difficulties were encountered in the Indian Ocean, due to a greater dispersion of continents and uncertain initial position of India and Madagascar. The Pacific region was the most difficult to reconstruct, as workers to follow also found. Unlike the Atlantic and Indian Oceans, where the borders retained their shapes, the Pacific borders were considered to have opened much earlier and remained tectonically mobile throughout the dispersal times.

To explain the expansion process Hilgenberg postulated that the mass of the Earth, as well as its volume, waxed with time. Because of this stance, and several problems inherent in his reconstruction however, Hilgenberg's ideas on Earth expansion were largely ignored and he received scant recognition for his efforts.

During the next 30 years the concept of an expanding Earth was advanced by Halm, Keindl, Egyed, Carey, and Heezen, developed primarily in the German and Russian literature.

Brösske: Used the present coastlines of the continents and assembled them onto a globe at 55% of the present Earth. His assembly resembled Hilgenberg however it had a tighter relationship for Africa, the Americas, Greenland, Europe and Asia. In closing the Pacific Ocean Brösske located eastern Australia into the Peru re-entrant, and the New Guinea front of the Australian plate against California, while the East Indies was packed between the Australian northwest shelf and Southeast Asian mainland.

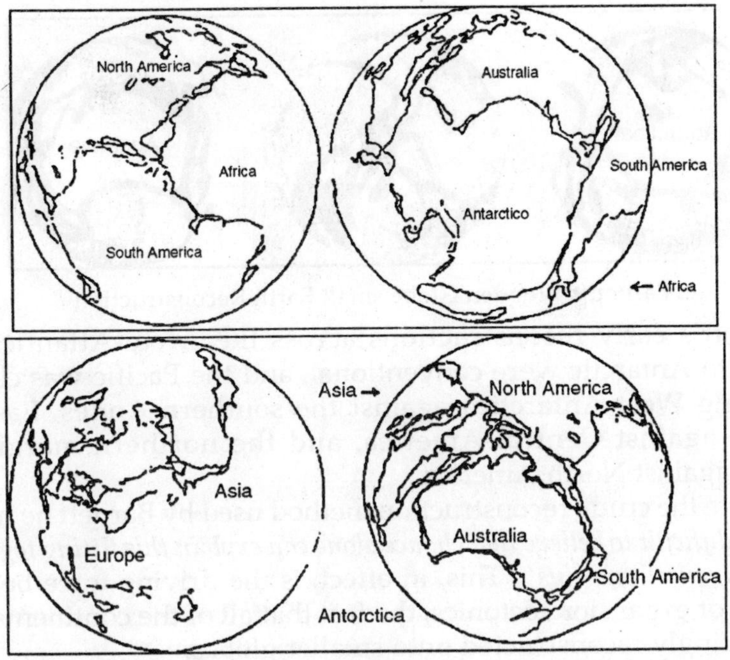

Fig. Brösske's Small Earth Reconstruction, at 55% of the Present Earth Radius, Showing a Tighter Relationship for Africa, the Americas, Greenland, Europe and Asia than Hilgenberg

Barnett: Adopted the 1000 fathom (200 metre) isobath for his reconstruction on a smaller globe as Hilgenberg had done. He used rubber templates cut from a 4½ inch diameter globe and reconstructed them on a 3 inch diameter wooden globe, representing a palaeoradius of about 65% of the present day Earth.

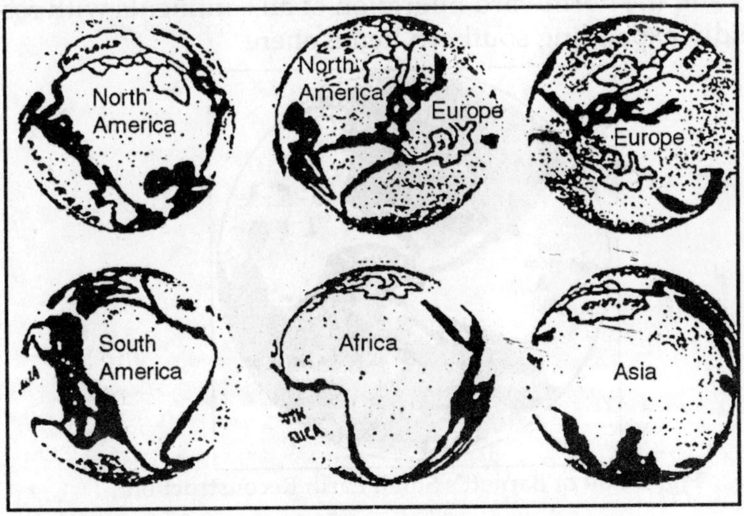

Fig. Barnett's Wooden Globe small Earth Reconstruction

Barnett's early reconstructions across the Arcto-Atlantic and Australia to Antarctic were conventional, and the Pacific was closed by bringing West Antarctica against the southern Andes, Eastern Australia against Central America, and the northern margin of Australia against North America.

Despite the crude reconstruction method used by Barnett he noted that, *"it is difficult to believe that chance alone can explain this fitting together of the continental margins"*. This, in effect, is the driving force behind advocates of expansion tectonics, the fact that all of the continents can be convincingly reconstructed onto smaller globes.

Barnett later made a new reconstruction bringing together the southern points of Africa, Australia, South America and the Antarctic peninsular to a common point near the Falkland Islands. He presumed this point to be the site of "primary rupture of the Earth". Carey similarly considered this point to be the centre of maximum dispersion for all the land-masses of the world. Barnett's models were the first to emphasise the Earth's hemihedral asymmetry, the antipodal relation of continents and oceans, the greater separation of the southern continents and the northward migration of all continents with respect to the rapidly expanding southern hemisphere.

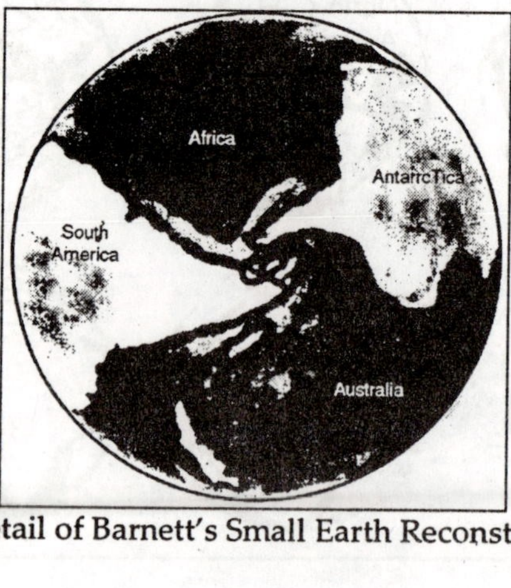

Fig. Detail of Barnett's Small Earth Reconstruction

Barnett also made comment on the newly explored world wide oceanic "rise-ridge systems" (mid oceanic rift zones), suggesting that, in the early phase of the Earth's history, all the tension rifts that gave rise to the oceanic "rise-ridge systems" were mid-oceanic, encircling the continental plates.

Creer carried out model experiments by enlarging fibreglass shell models of continents by a factor of 1.82 and assembling them onto a 50 centimetre globe, corresponding to a radius change of 55% of the present size Earth. The impression Creer received was that, "the fit of the continents on a smaller Earth appeared to be too good to be due to coincidence and required explaining". The fits obtained at 0.55, 0.77 and 1.0 of the present radius are shown in Figures.

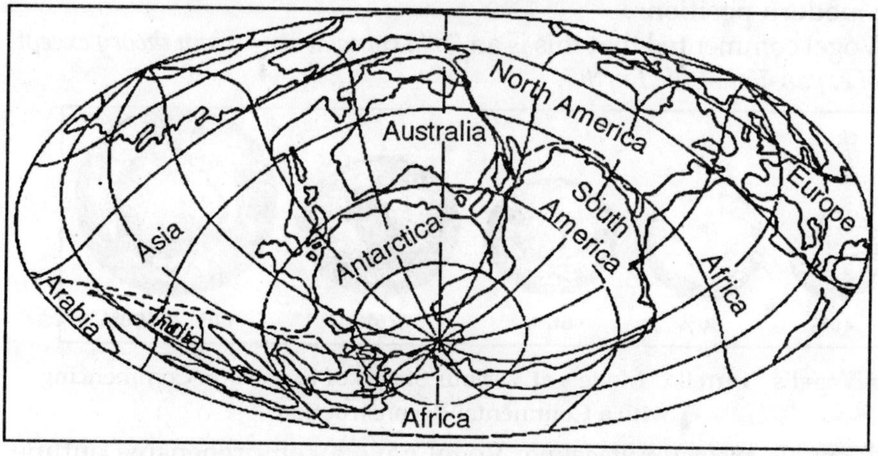

Fig. Creer's small Earth Reconstruction, at 55% of the Present Earth Radius

According to Creer the continental crust first developed a U-shaped crack between Australia, America and Asia during the "Early Precambrian". Subsequent Earth expansion was then largely taken up along this initial crack, which widened to form the Pacific Basin. His Atlantic assembly was similar to present day conventional reconstructions, however difficulties were still raised with the Indian Ocean. Creer included a Tethys which suffered an unconvincing discontinuity of more than 6000 kilometres from Saigon to its New Guinea continuation on the site of the Aleutians.

In contrast to other workers: Creer envisaged that expansion occurred early in the Precambrian. Being strongly influenced by the emerging conclusions of palaeomagneticians he concluded that the Earth's radius was not appreciably different from its present value when the fragmentation of Laurasia and Gondwanaland commenced during the Permian or Triassic.

Vogel: To date has produced the most comprehensive set of models, or "terrella" as he referred to them, at diameters of 40%, 60%, 66% and 75% of the present sized Earth plus a unique representation of a 55% reassembled globe inside a transparent plastic sphere of the present diameter.

Vogel's models again confirmed the greater dispersion of the southern continents, and he noted also a marked westward movement of all the northern continents relative to the southern continents, along what is now referred to as the "Tethyan shear", described by Carey. These models, based on mid-oceanic ridge systems and sea floor spreading zones, demonstrated to Vogel that, in general, the continents tended to move out radially from their Precambrian positions to reach their modern positions.

Vogel commented that this is an *"odd coincidence for any theory except that of expansion of the Earth"*.

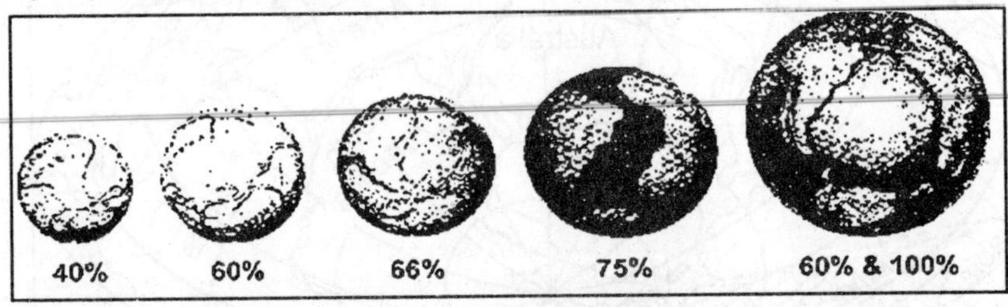

Fig. Vogel's "Terrella" Models at Various Stages of Expansion Commencing with a Continental Reconstruction

From his extensive modeling: Vogel gave a comprehensive outline of the fit of continental fragments under the headings of Gondwanaland and Laurasia.

Development of the oceans, with increasing Earth radius, was considered to have commenced during the Mesozoic, after final dislocation of the continental fragments, due to widening of the main mid-oceanic fracture zones. Vogel went on to consider the two hemispheres as complementary counterparts, with no Tethys Ocean or other "sphenochasms" required.

Vogel concluded from his modeling that: At a reduced Earth radius of 55% to 60% of the present,

- The continental outlines can be fitted together to form a closed crust;
- The positions of the different continents with respect to each other remain generally constant, with their separation caused by a "radial expansion of the Earth" and;
- The cause of the movements of continents resulted from an

accelerating increase in radius with time, in accordance with sea-floor spreading.

Vogel commented that "an accordance of these three phenomena cannot be accidental", but due to "processes operating from within the interior of the Earth resulting in Earth expansion".

In addition to the spherical models presented above a number of authors have presented graphical models worthy of note, based on various modeling techniques.

Dearnley used a reconstruction of: Precambrian orogenic belts in the Superior and Grenville règimes to deduce an expanding Earth model and proposed that the Earth's radius was 4400 kilometres (69%) at 2,750 million years, and 6000 kilometres (94%) at 650 million years. Dearnley regarded the major crustal features, in particular the orogenic fold-belts, as a direct crustal response to the underlying convection current activity of the mantle, and suggested a relatively steady rate of expansion commencing as far back as 4500 million years ago.

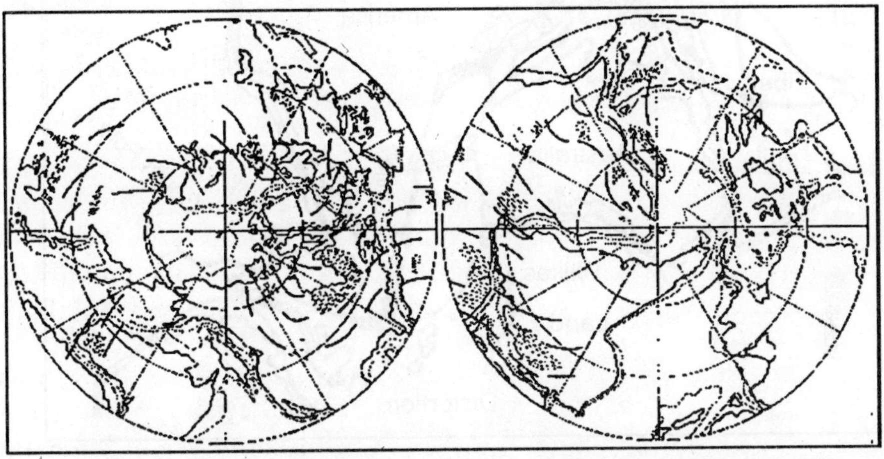

Fig. Dearnley's small Earth Reconstruction at a Palaeoradius of 4400 Kilometres

Shields: Based his model compilation primarily on palaeobiology, especially the trans-Pacific terrestrial biotic links that are not reflected in regions bordering the Atlantic. Shields used these biotic links to emphasis an east-west initial opening of the Pacific Ocean during Early Jurassic times, contrary to the plate tectonic view that the Pacific was formed from an even wider ocean during Mesozoic times.

Shields presented a new reconstruction of the continents bordering the Pacific, a new Arctic Ocean reconstruction, and a more or less conventional Atlantic and Indian Ocean assembly. He concluded that *"the Tethys Sea was much narrower than many suppose".*

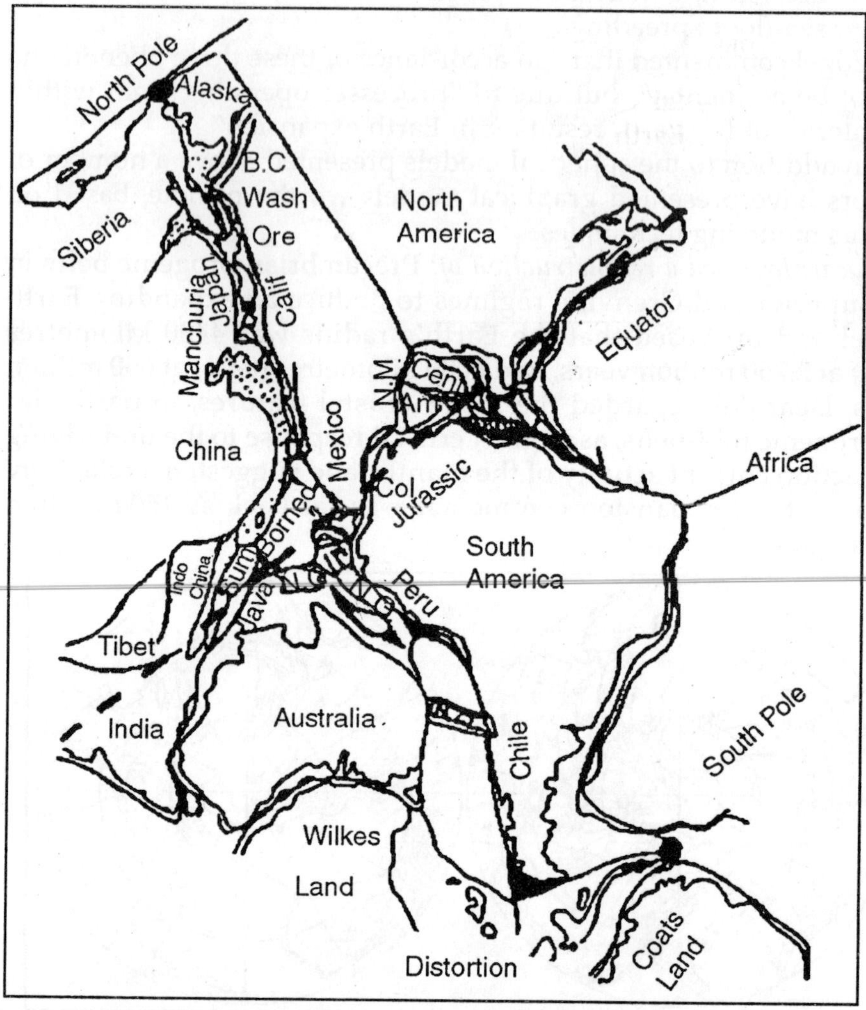

Fig. Right Shields'1976 small Earth Reconstruction of Continents Bordering the Pacific Ocean

Owen produced a voluminous atlas of continental displacement and expansion of the Earth during the Mesozoic and Cenozoic, based on a linear global expansion rate, commensurate with a 20% increase in diameter to the present mean Earth diameter, since the Late Triassic-Early Jurassic.

Owen adopted the 1000 metre isobath as the edge of the continental shelf and used base maps of the continents, cut from 2 millimetre thick expanded polystyrene foam, in conjunction with rubber inflatable spheres, to determine the point at which the various continents fitted together to reform Pangaea precisely. Using this method the diameter of the Earth in the Late Triassic-Early Jurassic was found to be approximately 80% of its modern mean diameter. The rate of Earth expansion was determined to be essentially linear for the past 200 million years. Although Owen concluded that reconstructions which assume a constant dimension Earth are untenable, he assumed that marginal subduction zones were active throughout the Mesozoic and Cenozoic, which is in direct contradiction to the Earth expansion views of Carey and Vogel.

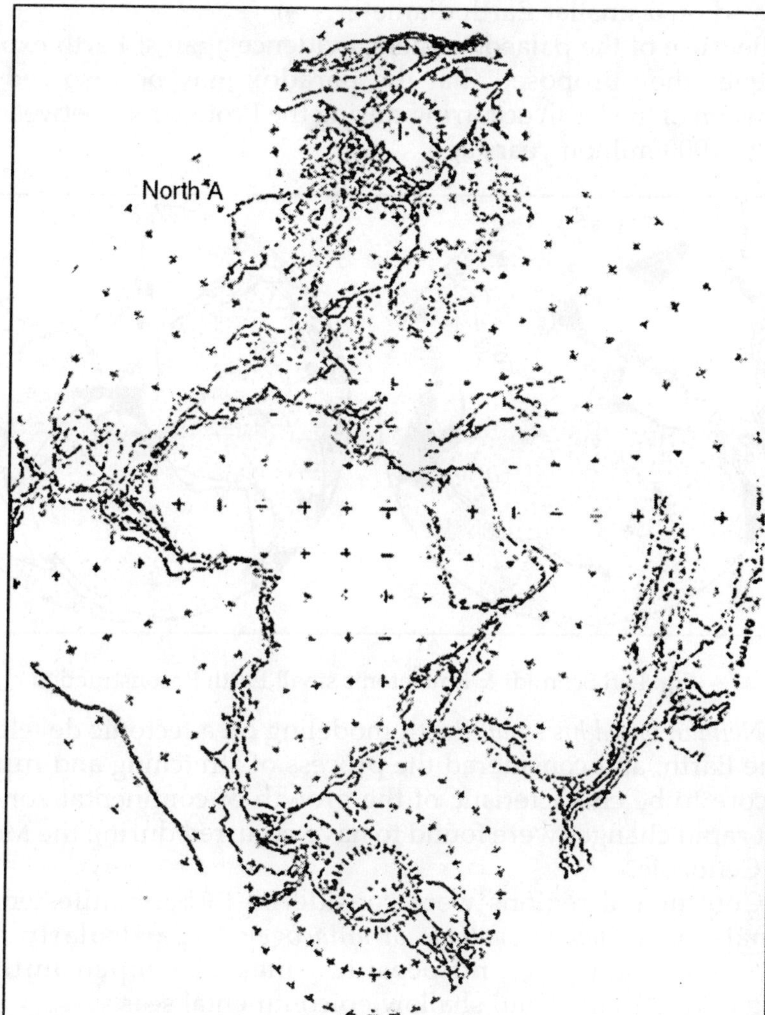

Fig. Right Owen's 1976 Small Earth Reconstruction of the Pangaean Hemisphere

Schmidt and Embleton: In commenting on the Early Proterozoic common apparent polar-wander paths for Africa, Australia, Greenland and North America deduced that there is abundant geochemical, geological, geochronological and tectonic evidence to suggest that these landmasses were much less dispersed in the Precambrian than they are now.

They used a small Earth globe, equal to 55% of the present diameter, to demonstrate that the Proterozoic geological, geochronological and palaeomagnetic information can be satisfactorily resolved on a smaller Earth diameter.

Because of the palaeomagnetic evidence against Earth expansion however they proposed that the paradox may be resolved by an expansion of the Earth occurring during the Proterozoic, between about 1600 to 1000 million years ago.

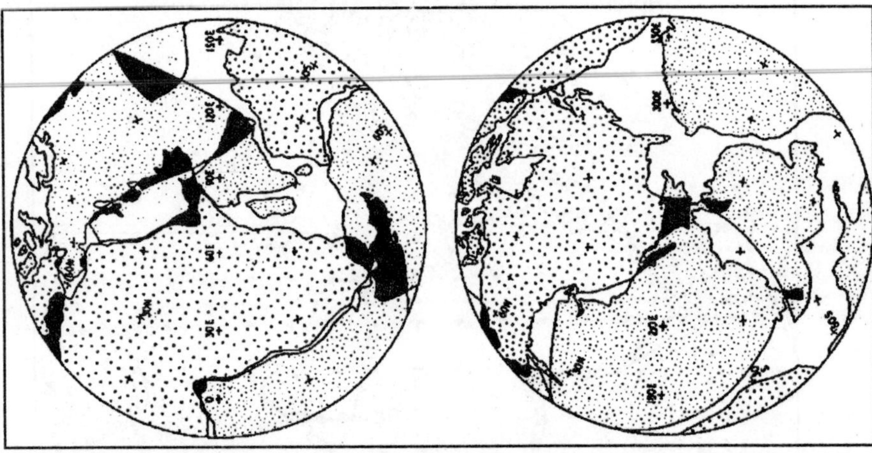

Fig. Left Schmidt & Embleton's small Earth Reconstruction

*Neiman based his small:*Earth modeling on a tectonic development of the Earth, and considered the process of stretching and rupture of the core to be characteristic of the growth of continental zones. The most rapid changes were found to have occurred during the Mesozoic and Cenozoic.

Continental regions were considered to be manifested in the formation of *"whole systems of aulacogens"*, particularly around continental margins, and oceanic zones developed initially as comparatively small and shallow epicontinental seas.

Neiman considered that the change in size of the: Earth varied from 16% of the modern radius at 2.2 billion years, 30% at 1.2 billion years, 37% at 1 billion years and 55% to 60% for the Mesozoic.

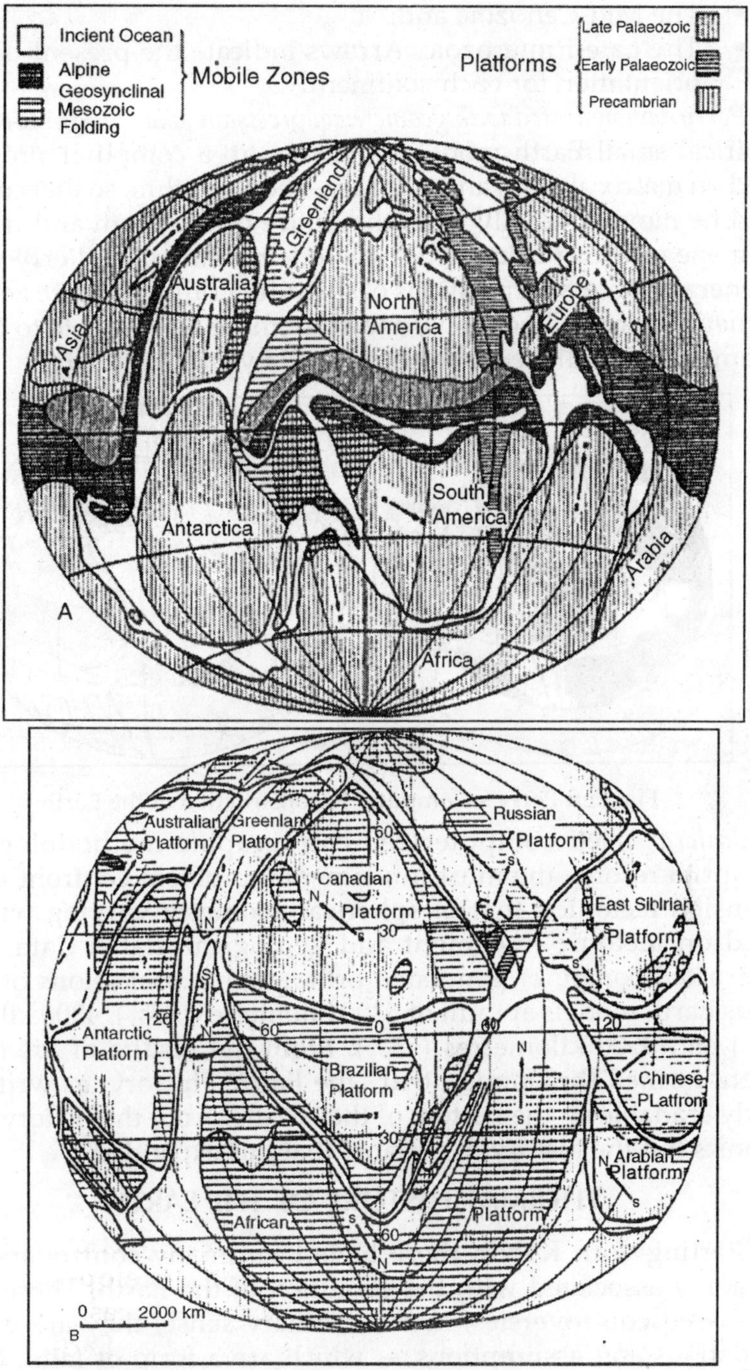

Fig. Right Neiman's Schematic Small Earth Tectonic Reconstruction of the Earth for

- The Mid-Cenozoic and,
- The Late Proterozoic. Arrows indicate the present-day north orientation for each continent.

Perry demonstrated with geometrical precision what: Vogel found from empirical small Earth modeling. He set up a computer programme based on matrix algebra and a hidden-line algorithm, so that continents could be moved radially from the centre of the Earth and translated using one centre of coincidence and one rotation pole. Perry was able to generate successive positions of spreading ridge, fracture zones, and magnetic anomaly lineations, and from these he was able to calculate the amount of radial expansion implied by each anomaly.

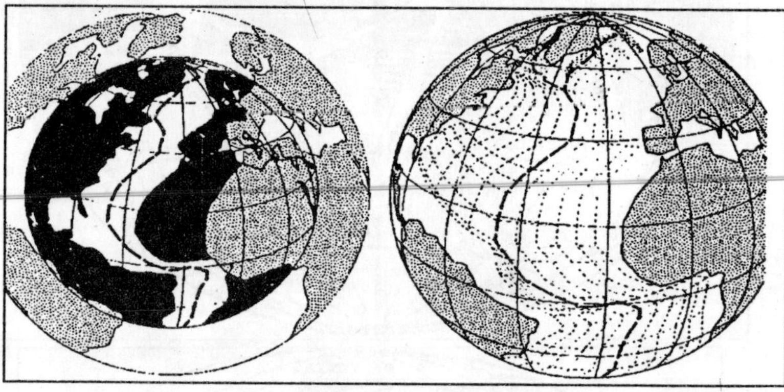

Fig. Left Perry's Computer Reconstruction of the Earth

Scalera: Used computerized cartographic methodology in an attempt to resolve the most fundamental problem confronting Earth expansion tectonics; that of spherical geometry varying with time. Based on tectonic lineament and geological-match data, Scalera constructed computer generated orthographic projections of smaller radius Earth models at radii of 5300 kilometres (83%), 4300 kilometres (67%) and 3500 kilometres (55%). From the results of his research Scalera formed the opinion that "the Earth's history, as written and clearly readable on the bottom of the oceans, is not the history of plate tectonics but the history of a planet that has expanded".

THERMAL HISTORY OF THE EARTH

Starting with Kelvin there have been many controversies and paradoxes associated with the evolution of the Earth. These are not faith-based controversies, in the ordinary sense; they are based on calculations and assumptions — which are a form of faith. Present-day heat flow is determined by the amount and distribution of radioactive elements, their secular decline with time, the delay between heat production and its appearance at the surface, secular cooling of the interior, and a variety ofminor sources of heating which are usually overlooked.

Chemical stratification of the mantle slows down the cooling of the Earth; the upward concentration of radioactive elements reduces the time between heat generation and surface heat flow. The initial conditions of the Earth cannot be ignored; they have not been forgotten. Continents divert mantle heat to the ocean basins and, in addition, tend to move toward cold downwellings, thereby protecting their cold keels. Surface conditions — including the thickness of plates — and mantle viscosity control convective vigour and cooling history of the mantle. A global accounting of the heat lost from the interior by hydrothermal processes at the surface is a lingering issue.

A chemically layered mantle with upward concentrations of the radioactive elements, shallow return flow, and a low-viscosity asthenosphere is the most plausible model of mantle dynamics. The migration of ridges and trenches is an important aspect of thermal history and geodynamics. The heat budget of the Earth cannot be treated as an instantaneous one-dimensional heat-flow problem, or one that involves a homogenous mantle with uniform and static boundary conditions. Both the radial and lateral structure of the Earth must be considered.

Viscosity of the mantle is temperature, pressure, composition, location and time-dependent. Magma oceans, plate tectonics, heat pipes and stagnant-lid convection have transferred heat out of the interior at various times. The current Earth approximates stagnant-lid conditions in one hemisphere and a lid that fully participates in internal convection in the other. On early Earth, the surface may have been covered with thick accumulations of unsubductable basalt, penetrated by pipes of magma, as appears to be the case on Io.

In such a scenario the rate of heat loss is regulated by the stress and density of the basalt pile as much as by the viscosity of the interior. The interior cools by depression of the cold surface layer and delamination, which displace hot material upwards. Even on today's planet, eruption and mantle cooling depend on the stress state of the overlying plate.

KELVIN

Lord Kelvin assumed that the Earth started as a molten ball and calculated that it cooled to its present condition by thermal conduction. Kelvin's assumed initial conditions are probably close to being correct. The energetics of accretion, core formation, giant impacts and formation of the Moon result in a hot, partially molten initial condition; gravitational stratification was unavoidable. Subsequent refinements of Kelvin's thermal history incorporated radioactivity and convection.

The acceptance of convection and the rough equality between the heat output of chondritic meteorites and early estimates of the total terrestrial heat flow led to the view that there is essentially a steady state in which heat production and heat loss are balanced. However, this violates the first law of thermodynamics; it also cannot be the case since radioactive heating of necessity decreases with time. The secular cooling of the Earth — the Kelvin effect — is important but has often been neglected or dismissed in recent years.

A perceived 'missing energy source,' controversies about 'corrections' to observed surface heat flow, origin of komatiite, and various Archean paradoxes and catastrophes are prominent in current discussions. The fact that there are no thermal or heat flow anomalies associated with hotspots is a paradox for the plume hypothesis [mantleplumes].

Current convection and geochemical models assume uniform radioactivity throughout the mantle, or a depleted upper mantle and a primordial or enriched lower mantle — these also cannot be correct. Global heat-flow estimates range from 30 to 44 TW. Estimates of the radioactive contribution from the mantle, based on cosmochemical considerations, vary from 19—31 TW. About 5—10 TW enters the mantle from the core. Thus, there is either a good balance between input and output, as once believed, or there is a serious missing heat-source problem, ranging up to a deficit of 25 TW.

Attempts to solve the perceived deficit problem by invoking secular cooling, or deeply-hidden heat sources — stealth layers in the mantle — have run into problems, some real, some perceived. Survival of ancient cratonic roots, komatiitic temperatures, an Archean heat flow catastrophe, overheating of the lower mantle, and helium heat flow paradoxes have been cited as potential problems. These concerns have replaced the *chondritic coincidence* that preoccupied geochemists when the first estimates of global heat flow came in, which suggested a steady-state Earth with an instantaneous balance of deep heat production with surface heat flow.

INITIAL CONDITIONS

A type of radial zone refining process accompanies the accretional process. This sweats out the crustal and radioactive elements and keeps them near the surface and drains the dense metallic melts and refractory crystals toward the interior. This chemical stratification, plus radioactive heating, stretches out the subsequent cooling of the mantle. Convection and the subduction part of plate tectonics accelerate the cooling but the surface and internal thermal-boundary layers are still conduction bottlenecks.

In the limit of multiple layers the Earth approaches the conduction cooling condition calculated by Lord Kelvin. At the high pressures and temperatures in the deep interior both lattice conductivity and radiative conductivity may be much higher than in the shallow mantle, again making the mantle approach Kelvin's assumptions. Under some conditions the Earth can forget its initial thermal state but chemical stratification and redistribution of radioactive elements cannot be forgotten. A chemically stratified mantle has a low effective Rayleigh number; vigorous convection and chaotic mixing are not expected.

In models of geodynamics and geochemistry that were popular in the last century, and are still in the textbooks, the lower mantle was assumed to have escaped accretional differentiation and to have retained primordial values of radioactivity and noble gases or even to have undetectable radioactively enriched layers. The crust was derived entirely from the upper mantle, making the latter extraordinarily depleted in the radioactive and volatile elements. However, mass balance calculations and the 40Ar content of the atmosphere show that most, if not all, of the mantle must have been processed and degassed in order to explain the concentrations of incompatible and volatile elements in the outer layers of the Earth.

The accretional zone-refining process results in an outer shell that contains most of the U, Th and K, at levels about three times chondritic (if the outer shell is equated with the volume of the present upper mantle), from which the protocrust and basaltic reservoirs were formed. The residual (current) upper mantle retains radioactive abundances greater than chondritic while the bulk of the mantle, including the lower mantle, is essentially barren. Previous chapters explored this possibility. The outer shells of Earth probably also contain, or contained, the bulk of the terrestrial inventory of noble gases. This is the reverse of the classical models and those used in convection simulations. In more recent models of geodynamics the entire mantle is assumed to have escaped chemical differentiation, except for crust extraction, and to convect as a unit, with material circulating freely from top to bottom.

This is one-layer or whole mantle convection. Recycled material is quickly stirred back into the whole mantle. In these models are:

- Chemical heterogeneities are embedded in a depleted matrix,
- The whole mantle is uniformly heterogenous

- The mantle is relatively cold and the hot thermal-boundary layer above the core plays an essential role in bringing heat to the surface.

Convection is assumed to be an effective homogenizer. An amendment to this idea is that there is a radioactive-rich layer deep in the mantle, but this is based on unlikely assumptions about upper-mantle radioactivity. Ancient radioactiverich regions in the deep mantle (if they survive accretional differentiation) will overheat, overturn and deliver their heat and heat-producing elements to the shallow mantle.

STYLE OF MANTLE CONVECTION

The deep mantle convects sluggishly because of the effects of pressure on thermal expansion, thermal conductivity and viscosity. Large grain size, high temperature and partitioning of iron may increase the ability of the lower mantle to transmit heat by conduction and radiation. The buoyancy flux of hotspots is often equated with the heat flux from the core but this heat may also be conducted readily into the base of the mantle and become part of the general background flux of the mantle. In any case the heat flow from the core and the heat generation of a depleted lower mantle are minor compared to the heat generation, secular cooling and plate insulation effects, and lateral temperature gradients in the upper mantle.

Plate tectonics, recycling and magmatism are forms of convection but the active layer may not extend deep into the mantle where the effects of pressure, low-heat sources andmore uniform thermal gradients suppress the importance of convection. The plate tectonic style of convection and heat removal may not extend too far back in time. At upper-mantle conditions, viscosity, conductivity and thermal expansion are strongly dependent on temperature. When this is so, mantle convective vigour adjusts itself to remove more heat as the temperature rises, unless the plates at the surface do not allow this. At high pressure, temperature is less effective in changing thermal and physical properties that depend on volume and there is less negative feedback. Viscosity also depends on water content.

Recycling serves to reintroduce water into the upper mantle, which promotes melting and a lowering of viscosity. One might think that a hotter Archean Earth convected more vigorously than the current mantle. But if melt and volatile removal are more important than recycling then a hot dry early mantle may have convected more slowly than the current mantle. If so, present-day secular cooling can contribute to the observed heat flow with no Archean thermal catastrophe. The ability of continents to divert heat to the ocean basins, and to move toward cold mantle, means that the stability of continental crust and mantle does not imply the absence of secular cooling for the mantle as a whole.

The mantle may have cooledmore slowly during the Archean than now because of the combined roles of subduction depth, recycling and mantle viscosity on mantle heat loss. The fact that the core is still mostly molten and the upper mantle is still near or above the melting point implies a long drawn-out cooling process. This favors a chemically stratified mantle but also one with most of the radioactive elements toward the surface.

The drawn-out cooling also implies that mantle viscosity is not just a function of temperature or that surface boundary conditions may also regulate the vigour and style of mantle convection. A chemically stratified mantle, even with the radioactivity toward the top, will cool slowly. On the other hand, if the deep layers are undepleted or enriched, they will warm up with time, and the system may overturn.

REGIONAL EARTH MODELS

The mantle loses two to three times as much heat per unit area in oceanic areas as in continental areas. One cannot discuss heat flow through continents, at present or in the past, without taking into account this partitioning. A corollary is that geological evidence from continents provides little constraint on average mantle temperatures or secular cooling rates. Continents affect the style of mantle convection, the rate of heat loss and the locations of downwellings.

The growth rate of continents and the partitioning of mantle heat loss below oceans and continents may serve to maximize mantle heat loss. The motions of continents, plate reorganizations and the creation of new plate boundaries permit pent-up heat to escape and these introduce temporal variability into global heat flow. The petrogenesis and inferred temperatures of Archean komatiites and the survival of cratonic roots may argue against an extremely hot mantle in the Archean and, thus, against substantial secular cooling.

It has been thought that such cooling could be ruled out since the very high temperatures and convective vigour predicted for the mantle do not show up in continental geology; this is the Archean paradox. These arguments are based on one-dimensional and static continent considerations. Continents both affect the distribution of heat flux and move around. A large fraction of the heat flow, and the evidence for possibly elevated mantle temperatures in the past, is therefore in the missing ocean basins and does not show up in the thermal history of continents. The Americas and Australia are currently over-riding mantle cooled by recent subduction.

The African crust and lithosphere have little knowledge of temperatures and convective vigour of the Pacific mantle. Realistic thermal history calculations include the coexistence of regional forms of convection and the multiple branches available to mantle evolution and cooling. Continents also introduce a long-wavelength lateral temperature gradient at the surface of the mantle. This alone can drive mantle flow. These lessons show us that secular cooling — the Kelvin effect — may not be lightly dismissed as an important contributor to surface heat flow, that the heat loss in oceans is affected by the presence of continents, and that mantle convection can be organized and driven by conditions at the surface. The plates may control the cooling of the mantle.

THERMAL HISTORY

There can be a long delay between the generation of heat in Earth's interior and its arrival at the surface. A tabulation of current heat sources and heat losses is therefore only part of the story. To understand the Earth's thermal budget requires information about thermal history. Some estimates of global heat flow appear to greatly exceed that available from current radioactive heating in the mantle. The ratio of heat production to total heat flux is called the Urey ratio.

The viscosity of the mantle is an important issue. Viscosity is temperature dependent and high temperatures result in low viscosity and increased convective vigour. This negative feedback serves to buffer mantle temperature. However, water content also affects viscosity, and the melting point. Magmatism extracts water from the mantle causing the viscosity to increase, even if the temperature is high. Therefore, high convective vigour and extensive magmatism in the past could actually lead to decreased convective vigour and lower heat flow. Magmatism plays a minor role in current estimates of heat flow but it may have been a more important mechanism for removing heat in the past.

The current plate-tectonic mode removes heat by plate tectonics (plate cooling at the surface and slab cooling in the interior). Io, a moon of Jupiter, may be an example where continuous volcanism, and the subsidence of the surface layer — the heatpipe mechanism — rather than plate tectonics, is the mechanism for coping with the prodigious tidal heating. The current style of plate tectonics also involves massive recycling of water into the interior. This means that the melting point of the mantle is low and the viscosity is low, in spite of relatively low temperatures compared to the Archean. Magmatism is primarily restricted to plate boundaries and regions of tensile stress; the latent heats of crystallization contribute much less than the heat associated with cooling from high temperatures.

This may not always have been the case. The cooling rate and cooling mechanism of the mantle are unlikely to have been the same throughout time. The vigour of mantle convection does not depend only on mantle temperature and viscosity. Plates, slabs and continents affect the style of mantle convection. Plates dissipate energy by colliding, bending, breaking and moving past each other and the underlying mantle. The buoyancy that drives convection can be dissipated in the plate system as well as in the mantle. Plates may actually regulate mantle convection and over-ride the buffering effect of mantle viscosity.

Thick plates, and a jammed plate system, can allow the mantle to over-heat, compared to the viscosityregulated effect. Thus, the arguments against high mantle temperatures in the past, the viscosities associated with such temperatures, the intuitive relationships between temperature and heat flow and the self-regulation of mantle temperature must be reconsidered [fate of fossil heat].

CHEMICALLY STRATIFIED MANTLE

We return once again to the possibility of a chemically stratified mantle. This possibility has apparently been ruled out by some workers based on visual inspection of a few colour tomographic cross-sections. One-layer or whole mantle convection is the current reigning paradigm in mantle geochemistry and geodynamics. The various paradoxes that plague studies of mantle dynamics and geochemistry can be traced to the assumptions that underlie the models.

In one class of models the mantle is divided into large reservoirs that can interchange, and exchange or entrain material. The deeper layers are highly radioactive and usually gas-rich. Slabs and plumes traverse the whole mantle, carrying the various components that are thought to characterize hotspot magmas. Ocean-island basalts (OIB) are assumed to come from the lower mantle, or the undegassed mantle. The upper-mantle reservoir is called 'the convecting mantle,' implying that it is well stirred by vigorous convection and is therefore homogenous.

Chemical stratification is the natural result of a planet growing in its own gravity field. The effect of pressure on thermal properties makes this stratification irreversible on a large planet. The standard models of mantle geochemistry and geodynamics, however, are quite different.

The reasons are not hard to find; unphysical assumptions have been used to rule out the possibility of chemical stratification and irreversible differentiation.

These include the Boussinesq approximation, assumption of a radioactive-rich lower mantle, assumed flat chemical interfaces between layers, and the assumption that the only plausible chemical boundary is the 650-km phase change. The textbook model of a depleted upper mantle, an undepleted, undegassed or primordial lower mantle, and a major isotopic boundary at 650 km, is unsatisfactory on many counts but this has been used to argue against all stratified models.

THE KELVIN EFFECT

The production of heat by decay of radioactive elements and the secular cooling of the mantle are the two main contributors to the present observed mean heat flux Q at the surface of the Earth. The first term is traditionally written as the Urey ratio, defined as the ratio between the heat production by radioactive elements and the total heat loss of the mantle. Attempts to reconstruct the thermal history of the Earth from a geophysical point of view have, since the time of Lord Kelvin, been in apparent disagreement with geochemical and geological observations.

The geophysical approach uses simplified parameterized models of mantle cooling, usually involving whole-mantle convection. Effects of time-dependence, sphericity, pressure, plates and continents are generally ignored. The rate of cooling in early Earth history obtained in these models is generally too rapid to allow a sufficient present-day secular cooling rate. Geochemical estimates of radioactive element concentrations in the mantle appear to be too low to explain the observed present mantle heat loss.

With present estimates of radioactive heat production in the mantle, simple parameterized models of wholemantle convection lead to a cooling of the Earth at the beginning of its history that is too fast and they are unable to explain the present heat loss of the Earth. This is one of the heat flow paradoxes of geophysics and it is related to the age of the Earth paradox of Lord Kelvin's time. There are several paradoxes associated with U, Th and K and their daughters such as heat, helium, lead and argon.

The balance between internal heat production and efficiency of heat transfer for a mantle with a temperature-dependent viscosity is such that the system is not sensitive to the initial conditions and self-regulates at each time step. A local increase in heat production or temperature results in a lower viscosity and more vigorous convection that carries away the excess heat. Mantle convection may not be self-regulating.

Cooling may be regulated by plate tectonics — involving the sizes and stiffnesses of the plates and the distribution of continents — leading to a weak dependence of the heat flow on the mantle viscosity and temperature. A hotter mantle results in a deeper onset of melting, more extensive melting, and a thicker buoyant crust and a dryer, stiffer lithosphere, which can reduce the total heat loss. Mantle viscosity and plate rheology depend on water content as well as temperature. These effects may be more important than the lowering of mantle viscosity by high temperature; a hotter mantle can actually result in lower heat flow.

It is common to assume the existence of a hidden mantle reservoir — a stealth or phantom reservoir — enriched in radiogenic elements and sequestered from global mantle circulation, to explain the present-day global heat budget. A deep, invisible undegassed mantle reservoir enriched in heat-producing elements has been the traditional explanation for the various heat flow and U—Pb—He paradoxes. But it is difficult to store heat-producing elements in the deep mantle. It is more likely that heat flow is variable with time, that plate tectonics controls the cooling rate of the mantle, and that the depleted MORBreservoir does not occupy the whole of the upper mantle.

There are various missing element paradoxes in geochemical mass-balance calculations and there is good evidence that the deeper mantle may be irreversibly stratified. But the inaccessible regions are more likely to be refractory and depleted. It is the assumption that depleted midocean ridge basalts and their depleted residues represent the entire upper mantle that is responsible for the idea that there are missing heatproducing elements. Recycled and delaminated crust, not currently at the Earth's surface, and ultra-enriched magmas, such as kimberlites, can readily make up the perceived He, Th and K deficits.

WRAP UP

One-dimensional and homogenous mantle models, or models with a downward increase in radioactive heating have dominated the attention of convection modelers. Paradoxes such as the Archean catastrophe, overheating of the lower mantle, persistence of cold continental keels and the missing heat-source 'problem' can be traced to these non-realistic assumptions and initial and boundary conditions. A variety of evidence indicates that high temperatures and efficient gravitational differentiation determined the initial conditions of the

Earth. On an Earth-sized body the effects of accretional energy and high pressure in the interior make chemical segregation essentially irreversible.

The outer parts of Earth contain most of the heat-producing elements and supply much of the secular-cooling part of the present heat flow. The major outstanding problems in the Earth's thermal budget and thermal history involve the role of hydrothermal circulation near the top, and the role of radiative transfer of heat near the bottom of the mantle. Convection modeling has not yet covered the parameter range that seems most pertinent from physical considerations and geophysical data. The largest failings in this regard are the neglect of pressure effects on material properties, and the use of parameterized convection.

What is needed is a thermodynamically self-consistent approach that includes the temperature, pressure and volume — dependence of physical properties, realistic initial and boundary conditions, and the ability to handle melting. Modeling has focused on models that are do-able, and which are perceived by modelers to represent constraints from other fields. These include whole-mantle convection or layered convection models with mass transfer between layers, the persistence of large isolated mantle reservoirs, and the need for vigorous chaotic stirring.

The perceived mismatch between heat sources and surface heat flow and the assumption of a homogenous upper mantle and undepleted lower mantle have led to a series of complex proposals regarding mantle overturns and depths of recycling. Continents affect the form of mantle heat loss; continents drift so as to be over cold downwellings or near trenches. The stability of cratonic roots and the temperatures of komatiites do not constrain the thermal evolution of ocean basins or the rate of mantle convection. Higher mantle temperatures may have led to more vigorous convection (the Tozer effect) but increased melting also removes volatiles and increases the viscosity and strength of the upper mantle-plate system.

The Hadean and Archean mantle may have been capped by a thick buoyant unsubductable layer and may have been drier and higher viscosity than the current mantle. There appears to be no mismatch between observed heat flow and plausible sources of heating. Current radioactive heating is just part of the equation. There are many open questions and opportunities for new approaches and ideas are enormous. The uncertainties are large and we must not make the same mistake as Lord Kelvin, who was confident in his estimate of the age of the Earth.

IMPLICATIONS FOR THE EARTH'S CORE

THERMAL EXPANSION OF IRON-RICH ALLOYS

The Earth's core makes up nearly one third of the planet's mass. Its composition, property, and dynamics are fundamental issues in the study of the Earth's interior. Deeply buried in the centre of the planet, the core has kept its chemical composition a long-standing mystery. Cosmochemical studies of meteorites and geochemical analysis of samples from shallower portions of the Earth suggest that the core is made of iron (Fe)-rich alloys containing nickel (Ni) and one or more lighter elements.

A critical test for any candidate core composition model is that it must be able to reproduce the physical properties of the core. On the basis of observing seismic rays penetrating the deep interior of the Earth and the orbital dynamics of the Earth as a planet in the solar system, models have been constructed to describe the physical state, density profile, and velocity profiles of the Earth's interior. One of the most widely used models is PREM. To perform the test of consistency between a composition model and the PREM model, we must know the thermal state of the core and the equation-of-state (EOS) of various Fe-rich alloys at the pressure and temperature conditions of the core.

The thermal state of the core can be deduced from the freezing point of core composition at the pressure of inner–outer core boundary and the adiabatic temperature gradient of the core composition under core pressures through laboratory measurements. Uncertainties in the core composition directly lead to uncertainties in the thermal state. By iteration, a self-consistent model of the thermal–chemical state of the core may be found. To complement this approach, we need accurate and precise phase relation and EOS data of various Fe-rich alloys at megabar pressures and high temperatures. In this paper, we review existing data on the thermal expansion of Fe-rich alloys and report experimental data on thermal expansion of Fe_3S, Fe, and Fe–Si alloys.

Thermal Expansion of Fe-Rich Alloys Under Static High Pressure

Thermal expansion is a fundamental aspect of EOS. Various parameters have been introduced to characterize thermal expansion under elevated pressures. They fall into two general categories, one focusing on macroscopic thermodynamic quantities and derivatives, and the other based on lattice vibration theories and microscopic view of solids. In the first category, *P–V–T* data are grouped into isothermal, isobaric, or isochoric sets. A basic approach is to calculate the thermal expansion coefficient from isobaric data according to its definition

$$\alpha = (\partial \ln V / \partial T)_P$$

The temperature dependence of thermal expansion coefficient is normally expressed in the form of $\alpha = a_1 + a_2 T$ or $\alpha = a_1 + a_2 T + a_3 T^{-2}$. One can also calculate temperature-dependent bulk modulus $(\partial K_{T,O} / \partial T)$ by fitting isothermal data to the high-temperature Birch–Murnaghan EOS. Isochoric data allow the calculation of thermal pressure at constant volume, which is related to thermal expansion coefficient and isothermal bulk modulus: $\Delta P_{thermal} = \alpha K_T \Delta T$,, where is αK_T commonly assumed to be constant at temperatures above the Debye temperature.

In the second category, the Mie– Gru¨ncisen–Debye EOS is widely used to extract a number of thermoelastic parameters from *P* – *V* – *T* data, including the isothermal bulk modulus at ambient pressure (*KT*, 0), its pressure derivative (*KT*, 0), the Debye temperature at ambient pressure and temperature θ_0, the Gru¨neisen parameter γo (which describes the volume dependence of the Debye temperature), and *q* (which is a parameter describing the volume dependence of the Gru¨neisen parameter). Jackson and Rigden carried out a systematic analysis of *P* – *V* – *T* data of mantle minerals and found excellent agreements between different methods. A survey of literature revealed a significant lack of thermal expansion data on Fe-rich alloys under static high pressure.

Extensive efforts have been focused on pure Fe. With the recent development in high-pressure and high-temperature experimental techniques and concurrent advance in synchrotron radiation facilities, this simplified core composition model has become amenable to investigations under increasingly higher pressures and temperatures. Using diamond–anvil cells and internal heating technique, Dubrovinsky *et al.*

determined the thermal expansion of Fe to a maximum pressure of 300 GPa and a maximum temperature of 1,500 K. Two multianvil studies covered similar range of temperature at moderate pressures. An independent estimate of the Debye temperature has been obtained from nuclear resonant inelastic scattering measurements, which determined the phonon density of states of Fe up to 153 GPa and at ambient temperature.

The Debye temperature at 1 bar derived from the phonon density of sta tes data agrees well with that from a x-ray diffraction study in a diamond– anvil cell (420–430 K), but they are in stark disagreement with the result of multianvil experiments (1,000 K). All of the reported thermal expansion coefficients at ambient conditions agree within uncertainties. Although laser-heated diamond anvil cells have been used extensively to determine the structural phase transition of Fe, no thermal expansion data have been reported. Due to limited temperature range covered by the existing static experiments, current discussions concerning the effect of thermal expansion on the density deficit in the Earth's core have largely relied upon shock wave data.

The leading candidates for alloying elements in the Earth's core include Ni, sulfur (S), silicon (Si), carbon (C), oxygen (O), and hydrogen (H). The phase relations and thermoelastic properties of a number of binary systems including Fe and one of the alloying elements have been studied under high pressures and temperatures. We found that thermal expansion data for Fe-rich alloys are extremely limited. With laser-heated diamond–anvil cells, thermal EOS of Fe_3S has been measured to 80 GPa and 2,500 K.

High-pressure data on Fe–Si alloys were collected in multianvil apparatus at pressures below 10 GPa. Two studies were conducted at ambient pressure using neutron diffraction method, one on Fe_3C and the other on FeSi. In general, the thermal expansion coefficients at ambient conditions are lower than that of pure Fe by 20%, with that of Fe_3C being the lowest. It is clear that more thermal expansion data are needed to evaluate the effects of alloying elements on the thermoelastic properties of Fe.

EXPERIMENTAL RESULTS AND DISCUSSION

Thermal Expansion of Fe_3S

In Fe-S binary system, Fe_3S with 16.1 wt.% S is the most Fe-rich sulfide known to date. Fe_3S belongs to the tetragonal crystal system with space group *I* 4 and is isostrucutral with Cr_3P. The unit-cell parameters of Fe_3S at ambient condition are $a = 9.144(2)$ and $c = 4.509(2)$ Å, with a zero-pressure density of 7.033 g/cm^3. Our x-ray diffraction patterns are consistent with this structure model, indicating no structural phase transition up to 42.5 GPa and 900 K. The volume-pressure data of Fe_3S at temperatures of 300, 600, and 900 K have been determined under pressures up to 42.5 GPa. The hightemperature Birch–Murnaghan EOS is used to fit the isothermal compressional data

$$P(V) = \frac{3}{2} K_{T,O} \left[\left(\frac{V_O}{V} \right)^{\frac{7}{3}} - \left(\frac{V_O}{V} \right)^{\frac{5}{3}} \right]$$

$$\times\left\{1-\frac{3}{4}(4-K_{T,O'})\left[\left(\frac{V_o}{V}\right)^{\frac{2}{3}}-1\right]\right\}$$

where

$$K_{T,O}\,K'_{T,O,}\,\text{and}\,Vo$$

and *V*0 are the bulk modulus, its pressure derivative, and the unit cell volume at zero pressure and temperature *T*, respectively.

The room-pressure unit cell volume *V*0 is given by the following expression: *V*0(*T*) =*V*0 (300 K) $\exp\left(\int_{300\alpha T,0}^{T} dT\right)$ $\text{where} V_o \text{and} \alpha_{T,0}$ are the unit cell volume at room pressure and 300 K and the thermal expansion coefficient at room pressure and temperature *T*, respectively. Results of the least-squares fit of the third-order Birch–Murnaghan EOS to the compression data are plotted in Figure. The *f F* plot confirmed that the compression data of Fe_3S are adequately described by a third-order truncation of EOS. The temperature dependence of isothermal bulk modulus $(\partial K_{TO}/\partial T)_P$ is estimated at. $-3.75\mp1.80\times10H^{-2}GPaK^{-1}$.

We calculated the thermal expansion coefficients of Fe_3S at 1 bar. Assuming a linear temperature dependence in the form of

$$\alpha_{T,O}=a_i+a_2T, \text{we Found}\, a_1=3.0\pm1.3\times10^{-5}K^{-1}$$

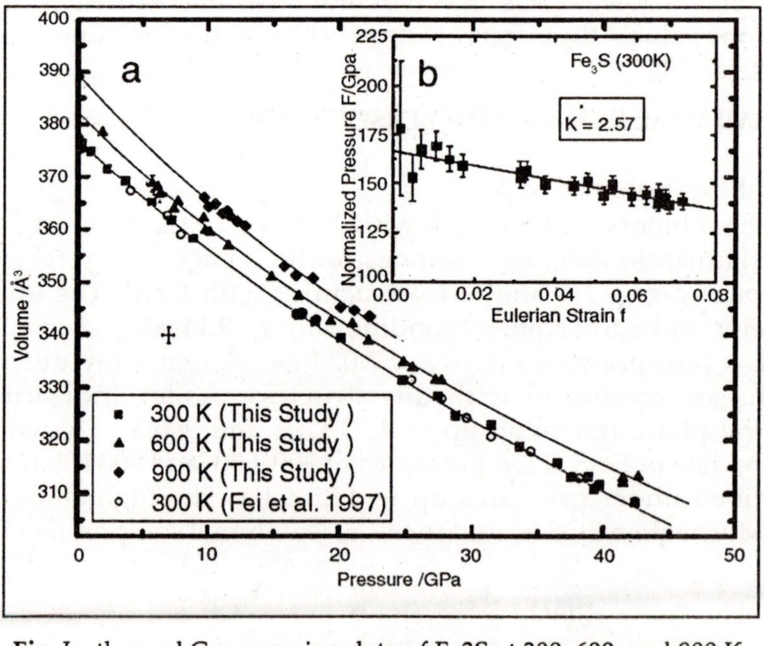

Fig. Isothermal Compression data of Fe3S at 300, 600, and 900 K.

- Open Circles Represent Compression data at room Temperature (300 K) from Fei *et al.*Filled squares, triangles, and diamonds represent the compression data at 300, 600, and 900 K, respectively. The three lines are the third-order Birch–Murnaghan EOS fits to the compression data. The error bars are estimated from multiple x-ray measurements at the same condition. Where duplicate measurements were not made, uncertainties estimated from multiple diffraction lines in a single measurement are slightly smaller than the symbols. A typical error bar is shown in the lower left corner.
- *f* - *F* plot of the compression data at 300 K. The data can be fitted by an inclined straight line, suggesting that the compression of Fe_3S can be adequately described by a third-order Birch–Murnaghan EOS. The error bars are based on multiple diffraction lines in a single measurement.

Fei *et al.* determined the compression curve of Fe_3S at 300 K. They reported isothermal bulk modulus *KT,0* = 170 GPa with a corresponding pressure derivative $K'_{T,O} = 2.6(5)$ or $K_{TO} = 150$ GPa 150 GPa with fixed $K'_{T,O} = 4$ Seagle *et al.* extended the thermal EOS of Fe_3S to 80 GPa and 2,500 K and determined the bulk modulus of Fe_3S at room temperature $K_{T,0} = 156$ GPa with pressure derivative $K_{T,0} = 3.8$ by fitting a third-order Birch–Murnaghan EOS to room temperature compression data by using NaCl in B2 structure as a pressure scale.

They also reported $\alpha K_T = 0.011(2)\,\mathrm{GPa\,K^{-1}}$ for Fe_3S, by expressing the thermal contribution of Fe_3S in the form of $\Delta P_{thermal} = \alpha K_T \Delta T$. The α_0 calculatedfrom

$\alpha K_T = 0.011(2)\,\mathrm{GPa\,K^{-1}}$ and $K_{TO} = 113$ given in Seagle *et al.* is significantly higher than our results. The discrepancy may result from the temperature dependence of thermal expansion coefficient. Extrapolating our calculated thermal pressures at 600 and 900 K to higher temperatures, our trend agrees with that in Seagle *et al.* within experimental uncertainties, but falls within the negative side of their error bars.

THERMAL EXPANSION OF FE AND FE–SI ALLOYS.

Pairs of x-ray diffraction measurements were taken on both pure Fe and an FeO.SiO. sample in the same cell assembly, within the pressure range of 21–27 GPa and the temperature range of 300 to >2,000 K. Within this pressure range and between 300 and 1,273 K, Fe adopts the hexagonal close-packed (hcp) structure, whereas FeO.864SiO.136

breaks down to a mixture of a Si-enriched bodycentered cubic phase and a Si-depleted hcp phase. These are proposed to be the stable forms of Fe and Fe–Si alloy in the Earth's inner core.

At higher temperatures, the hcp–Fe phase disappears; thus, data collected at temperatures above 1,273 K are not reported here. Calculated minimum density deficit of hcp–Fe0.Si0. relative to hcp–Fe from the paired data are shown in Figure. Because the hcp phase of Fe–Si alloy contains less Si than the starting material, these values represent lower limits of density deficits caused by 13.6 at. % Si. The variations in density deficit reflect varying Si content in the Fe–Si alloys as a function of pressure and temperature and the differences in compressibility and thermal expansion between the Fe–Si alloys and pure Fe.

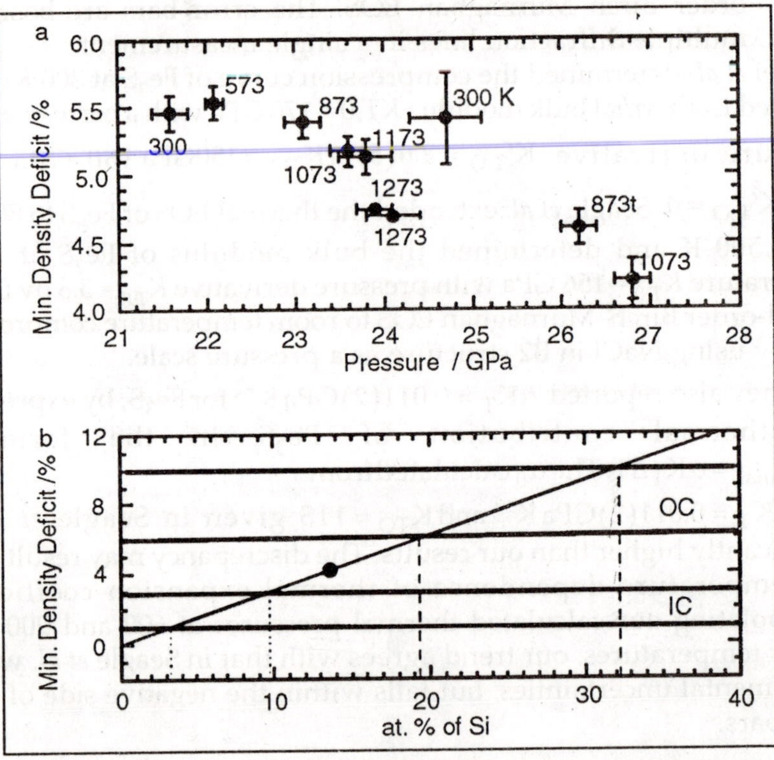

Fig. Density Deficits in Fe–Si Alloys and Si Content in the Earth's core.

- Minimum density deficit between hcp–FeO.864SiO.136 and hcp–Fe under different experimental conditions. Density deficit is defined as 100%, where represents the density. These data are listed in SI Table. Numbers next to symbols denote temperatures in Kelvin.

- Density deficit versus maximum Si content in Fe–Si alloy with hcp structure. Gray regions labeled "OC" and "IC" indicate density deficit of the outer and inner core compared with pure Fe, respectively. The filled circle represents the minimum density deficit produced by 13.6 at.%Si, based on data in *a*. The solid line represents a linear relationship between density deficit and Si content; dotted and dashed lines enclose the ranges ofmaximum Si content in the IC and OC, respectively.

Most existing EOS measurements are performed on individual phases for the sake of minimizing interference in x-ray diffraction spectra and possible chemical reactions. Any discrepancies in pressure and temperature between separate experiments would introduce uncertainties to the calculated density deficit. Direct comparison of densities between an Fe-rich alloy and pure Fe under identical conditions has the advantage of eliminating or reducing such uncertainties.

The technique of measuring symmetrically placed samples in a multianvil apparatus was pioneered by Zhang and Guyot. That of measuring well mixed samples in a laser-heated diamond–anvil cell was first applied by Seagle *et al*.

In this study, we have extended the pressure range of such multianvil investigation from <10 GPa to nearly 30 GPa. Our data place a lower limit on the capability of Si to reduce the density of solid Fe. Further constraints can be obtained by carrying out chemical analysis of run products quenched from each temperature and pressure conditions.

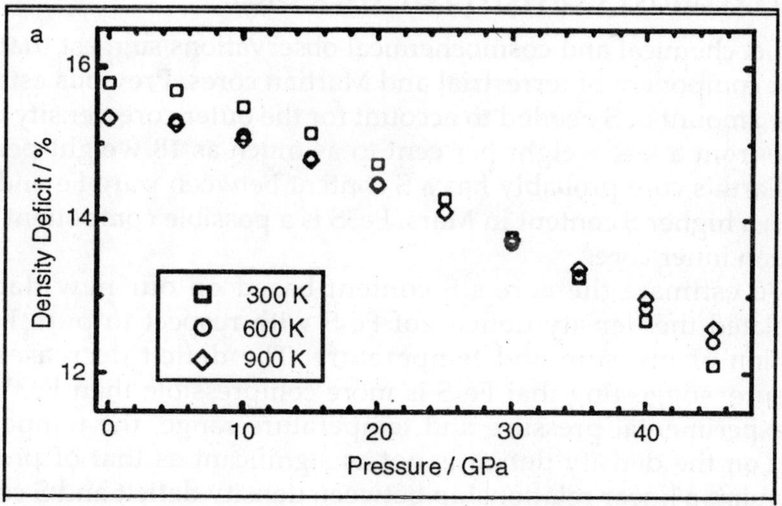

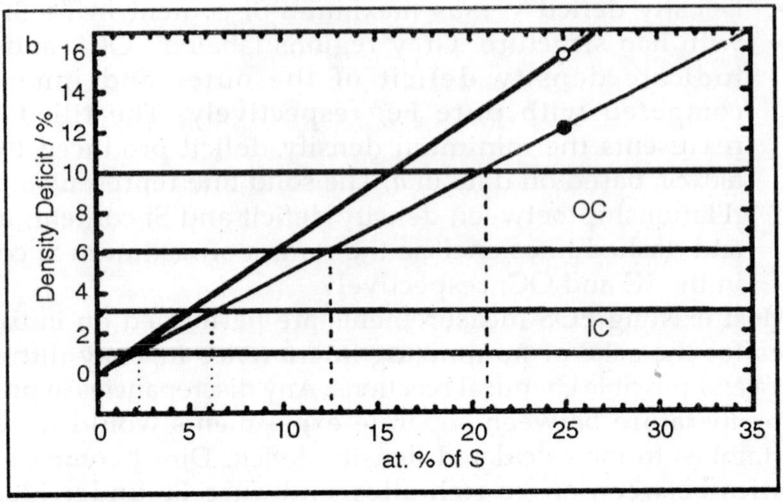

Fig. Density Deficit in Fe_3S and S Content in the Earth's Core

- Density deficit between Fe_3S and Fe, versus pressure.
- Density deficit versus S content in an Fe–S alloy. The open circle denotes the density deficits of Fe_3S with respect to Fe at 1 bar and 300 K. The filled circle denotes the deficits at 42.5 GPa and 900 K. Gray regions labeled "OC" and "IC" indicate density deficits of the outer and inner core with respect to pure Fe, respectively. The solid lines represent linear relationships between density deficit and S content; dotted and dashed lines enclose the ranges of S contents in the IC and OC, respectively.

LIGHT ELEMENT CONTENTS OF THE CORE.

Geochemical and cosmochemical observations suggest that S is a major component of terrestrial and Martian cores. Previous estimates of the amount of S needed to account for the outer core density deficit range from a few weight per cent to as much as 18 weight per cent. The Earth's core probably has a S content between pure Fe and Fe_3S. Given a higher S content in Mars, Fe3S is a possible component of the Martian inner core.

To estimate the core's S content based on our new data, we calculated the density deficit zof Fe_3S with respect to pure Fe as a function of pressure and temperature. The deficit decreases with pressure, suggesting that Fe_3S is more compressible than Fe. Within the experimental pressure and temperature range, the temperature effect on the density deficit is not as significant as that of pressure. Assuming a linear relationship between density deficit and S content, we estimated that at ambient conditions 1.6 at. % (0.9 wt. %) S can produce 1% density deficit. At 45 GPa and 900 K, 2.1 at. % (1.2 wt. %) S is needed to produce the same amount of density deficit.

Extrapolating our data based on Birch– Murnaghan EOS to the core conditions leads to unreasonable results, probably due to uncertainties in the fitted *KT* values. Given 6–10% deficit between the outer core and solid Fe, and 1–3% deficit between the inner core and solid Fe, we estimated 12.5–20.7 at. % (7.6–13.0 wt. %) S in the outer core and 2.2–6.2 at. % (1.3–3.7 wt. %) S in the inner core based on our data at the highest experimental pressure and temperature. Implicit in the estimated S content in the outer core is the assumption that S has a negligible effect on the volume change across the liquid–solid boundary.

Si is among the top three most abundant elements in the Earth's crust and mantle. A considerable amount of Si may be incorporated into the core through chemical reactions, especially under reducing conditions.

Geochemical data suggest that, depending on the accretion history and the distribution of Si in various layers inside the Earth, the Si content of the core may vary between virtually zero to 14 wt.%. Existing estimates of the amount of Si needed to account for 1% density deficit in the core range from 2.9 at. % (1.5 wt. %) to 5.8 at. % (3 wt. %). Assuming a linear relationship between density deficit and Si content, we estimated that, between 22.1 and 26.8 GPa, no more than 3.2 at. % (1.6 wt. %) Si is needed to produce 1% density deficit in hcp–Fe. This estimate agrees well with the lower estimate in the literature. Given 6–10% deficit between the outer core and solid Fe, and 1–3% deficit between the inner core and solid Fe, our experimental data suggest an upper limit of 19–32 at. % (11–19 wt. %) Si in the outer core, and an upper limit of 3.2–9.6 at. % (1.6–5.1 wt. %) Si in the inner core.

VELOCITY GRADIENT IN THE OUTER CORE

The fitting of our thermal EOS data to Fe_3S third-order Birch-Murnaghan EOS gives pressure derivative of bulk modulus *KT* < 4, similar to the results of Fei *et al.* Williams and Knittle proposed that *KT* can place an effective constraint on the primary alloying element in the outer core. They concluded that a *KT* value that is < 4 cannot reproduce the large gradient of bulk sound velocity in the outer core. Although their application of this constraint to Si was invalidated by later studies, it is still a useful criterion for testing core composition models. If the small *KT* of Fe_3S persists to higher pressures and temperatures, the model of a S-rich core would be challenged. The *KT* from our study and Fei *et al.* are much smaller than that given in Seagle *et al.* Independent studies are needed to resolve the discrepancy.

EFFECT OF SPIN TRANSITION.

Fe-rich alloys may undergo pressureinduced high-spin to low-spin transitions. Magnetic transitions in pure Fe and troilite (FeS) are accompanied by structural phase transformation, whereas those in Fe_3S and cementite (Fe_3C) do not seem to involve major structural changes. *Ab initio* calculations of Fe_3C predicted abrupt volume reduction at the transition, which has not been observed in experiments.

Nevertheless, such transition has been shown to affect the elastic and thermodynamic properties of Fe_3C and Fe_3S. Without prior knowledge of the magnetic transition, fitted EOS may be erroneous. At room temperature, the high- to low-spin transition in Fe3S occurs between 20 and 25 GPa. Concurrent reduction in the unit-cell volume was not observed in our study.

However, a plot of *c*/*a* ratio as a function of pressure revealed an obvious change between 20 and 27 GPa, with a valley located at 18 GPa. This pressure range matches that of magnetic transition in Fe_3S. The change in *c*/*a* ratio results from an increased resistance along the *c* axis accompanied by a softened *a* axis, with little change in volume. The data at 600 K suggest a similar but less obvious change in the *c*/*a* ratio between 25 and 28 GPa. The existence and condition of this change is poorly constrained due to the sparse data coverage.

The observed *c*/*a* ratio change may be associated with the high- to low-spin transition of Fe_3S, unless it is fortuitously caused by redistribution of stress due to the B1–B2 transition in the NaCl pressure medium, which occurs within the same pressure range. To determine the spin state of Fe in a S-rich core, we need more measurements to cover a larger pressure range at high temperatures. Different pressure media should be used to avoid the interference from phase transition in NaCl.

MATERIALS AND METHODS

Fe_3S. The starting material Fe_3S was synthesized stoichiometrically from metallic Fe and troilite (FeS) in a subsolidus experiment at 21 GPa and 1,000°C in a multianvil apparatus at Carnegie Institution of Washington. Electron microprobe and powder x-ray diffraction analysis indicate that the quenched run product is predominantly Fe_3S phase with <2% metallic Fe.

Energy-dispersive x-ray diffraction measurements were carried out by using polychromatic (white) wiggler synchrotron x-ray radiation at X17C beamline, the National Synchrotron Light Source, Brookhaven National Laboratory. The experimental procedure was similar to that described by Fei *et al.*

A germanium solid-state detector was used to collect the diffraction data at a fixed angle of 12.01°, which was calibrated by the x-ray diffraction spectrum of gold at ambient conditions. Two diamond–anvil cell assemblies were used for the measurements, both containing Fe_3S sample sandwiched between NaCl pressure medium. A small amount of MgO, Au, and Pt was placed next to the sample in one assembly, whereas only Au was included in the other. An externally heated diamond–anvil cell was used to generate high pressures and temperatures.

X-ray diffraction data were collected in several compression–heating–quench cycles covering pressures up to 42.5 GPa at 300, 600, and 900 K. Temperatures were measured with a platinum-rhodium (Pt-Rh) thermo-couple. The tetragonal unit-cell parameters for Fe_3S were fitted by using the GSAS/EXPGUI programme, primarily from the diffraction lines 321, 330, and 112. Typical diffraction patterns collected in this study are shown in SI Figure. Pressures were primarily determined from Au pressure scale. NaCl was also used as a secondary pressure scale. Above the B1–B2 transition in NaCl at 29 GPa, pressures were calculated only from the lattice parameters of Au.

FE AND FEO.SIO

The experiments on Fe and FeO.SiO. (7.3 wt.% of Si) were carried out by using a Kawai-type multianvil press at the BLO4B1 high-pressure and high-temperature beam line at the SPring-8 synchrotron facility in Japan. The experimental procedure was similar to that described by Fei *et al.*. In each run, we placed two samples symmetrically inside a cylindrical heater. One sample is a mixture of polycrystalline Fe (a few microns in average grain size, and >99.99% purity) and MgO powder (99.99% purity) at a weight ratio of 1:4. MgO served as a pressure marker, and was also used to inhibit Fe crystal growth at high temperature. The other sample is a polycrystalline FeO.SiO. alloy from Goodfellow.

The two samples were separated by a layer of packed MgO powder. In the first run, samples were first compressed to a target pressure of 800 tons (21 GPa). They were then annealed at 1,173 K to reduce nonhydrostatic stress, followed by a series of x-ray diffraction measurements along a heating ramp from room temperature to the peak temperature of 2,373 K. In the second run, a target pressure of

1,200 tons (25 GPa) and a peak temperature of 2,273 K were reached. Samples were annealed at 973 K. For the energy dispersive x-ray diffraction measurements, the solid state detector was fixed at an angle of 5.814° relative to the incident x-ray beam. The acquisition time for each spectrum is 200 s.

The position of x-ray beam was monitored through an x-ray absorption image. X-ray diffraction spectra were analyzed by using the XRAYANA software provided by the beamline, and using the GSAS/EXPGUI.

Chapter 8

Electrical State of the Atmosphere

It is known now from the phenomena of wireless transmission, that the very high atmosphere is greatly ionized, and that the amount and distribution of this ionization widely change between day and night. The telling of what these changes are, so far as known, explaining how they are measured, and speculating on their causes, make fascinating reading for the specialists who are engaged in, or at least understand, that kind of investigation. But the story is too long to attempt here, and besides would be out of date before it could reach the reader, so rapidly is it being developed.

It seems to be pretty well established, howover, from the phenomena associated with wireless transmission, that the upper atmosphere is much more strongly ionized during the daytime than it is at night, and more strongly when sunspots are numerous than when they are few. Furthermore, this ionization is largely concentrated at certain levels, especially at about 73 miles and 160 miles, respectively, above the earth. This upper layer, however, appears to be quite variable in height and often even to divide into two layers of considerably different heights. Such are some of the facts of observation, and there are many others, but how they are related, one to another, and exactly how they are caused we do not yet know, and may never know fully, though day by day we shall surely come nearer to that complete understanding.

We do not know, for instance how, nor from what source, the earth gets her continual vast charge of negative electricity despite a steady out-current of 1,000 amperes. There is good reason to believe that this state of the earth's surface is largely owing in some way to the action of thunderstorms; partly to emanations from radioactive elements; and partly, also to cosmic rays from "Lord-knows where"; and partly, perhaps, to many other sources. We know, furthermore, that the ever-flowing air-earth electric current, if undiminished, would completely discharge the entire earth in 7.5 minutes. Whence, then, comes this current, ceaseless as the fall of a cataract that never runs dry?

SAINT ELMO'S FIRE

Nothing is more awesome and soulenthralling than the unbelievable realities of a dark and stormy night on a tropical sea. Beneath, and all around, a vast and glowing fire that consumes not nor is consumed; and above, on every higher point, a cold, ghostly flame of sulphurous odor.

The first of these mysteries, occurring, as it does, beneath the water surface, is not in any sense a manifestation of atmospheric electricity, and therefore, no discussion of it will be offered here. True it is likely to be exceedingly active whenever the tipping flames occur, but not owing to any causal relation between them, for they are of wholly independent origins—it is just that a storm intensifies the one and may induce the other, and that the two together cast a deeper spell than could either alone of awesome mystery.

The flames, however, that blaze without burning on the topmost points, or jump from place to place like torches thrown by invisible hands are wholly electrical—just one of the beautiful things the physicist loves to play with in his toy house, the laboratory. There he sets a suitable electric machine running in the pitch dark, and presto it is all ablaze with bushy flames, "brush discharges," he calls them, from a hundred places. Tufts of frazzled lightning is what they really are, even if that term does seem a trifle disrespectful of Jove's mighty thunderbolts. The same phenomenon appears also along certain power lines that carry deadly high-potential currents. In this case the glow caused by the loss of electricity from the wire into the air is called a corona, and the loss itself a coronal discharge.

But "brush discharge" and "coronal discharge" are recent technical terms used by the physicist and the engineer in speaking of the loss into the atmosphere of electricity from objects charged to such high potential that the resistance of the air breaks down under the strain. Exactly this same phenomenon, caused in the same way—a brush discharge due to great electric stress (high voltage) occurs from time to time, and especially during the passage of a thunderstorm, on the high points of ships at sea, and on land from tall towers and particularly from the tops of isolated peaks.

THE AURORA

To one who had never seen nor heard of such a thing a glowing red midnight sky might portend the crack of doom; and even more appalling would be a sky hung with luminous draperies, fold upon fold; and terrifying in the extreme a forest of gigantic flaming swords, furiously slashing across half the heavens. And yet to all this, and more, the Esquimo pays not the slightest heed. To him the aurora in its many forms and various colors is a familiar sight.

We are certain he has no idea what it is, for he explains it in much the same way that we did, not so long ago, but he does know that it is neither harmful itself nor the harbinger of evil, and so to that which in many another would inspire awe, at least, and perhaps even terror, he gives not the remotest thought.

Familiarity begets indifference. As just implied, the aurora is a night-time glow in the sky that more or less resembles the light produced by an electric discharge through a rarefied gas. No doubt it occurs in the daytime quite as often, on the average, and as brilliantly as at night, but it is not then visible because the brightness of the day-sky is so much greater than that of the aurora that the glow of the latter is completely lost in the glare of the former. Neither can it be seen when the sky is completely overcast, because its closest approach to the surface of the earth still is far beyond the loftiest cloud.

In respect to region of frequent occurrence there are two classes of auroras, called, respectively, northern lights, northern aurora, or aurora borealis; and southern lights, southern aurora, or aurora australis, according as reference is had to its appearance in the northern or southern hemisphere. When, however, reference is to the shape of the luminous region, as seen against the heavens, auroras may be divided into arcs, bands, rays, curtains, coronas, patches, diffuse glows, and as many more as one's fancy can suggest and his vocabulary supply—mostly useless.

In the northern hemisphere the aurora is most frequent at about latitude 60° over North America and the Atlantic Ocean, and near 70° in Siberia. In short, the locus of its most frequent appearance in the northern hemisphere is, roughly, a great elliptical ring around the north magnetic pole, but much nearer to it on the American side than the Asiatic. Away from this ring the occurrence of the aurora is rapidly rarer with decrease of latitude. In the region of the Great Lakes, for instance, as in Scandinavia, it is quite common; unusual in the Gulf States and along the Mediterranean; and rarely seen within the Tropics.

As already stated the aurora can not be seen against the glare of a sun-lit sky. Hence, to the eve it is strictly a night phenomenon, like the Milky Way, and for the same reason it is more frequently observed during the winter, when the nights are long, than in the summer time when they are short. In addition to these diurnal and annual periods, of obvious cause, in the appearance of the aurora, it also has an eleven-year period, coincident with the sunspot period in the sense that when the spots are most numerous the auroras are most frequent, and rarest when the spots are fewest.

It, therefore, parallels also, at least roughly, the occurrence of magnetic storms. It has no relation to season, beyond that incident to the length of the night; none to the phases of the moon except that faint auroras cannot be seen in bright moonlight; nor any relation to the weather, except the obvious one that it never is seen when the sky is overcast.

Evidently not, for clouds never rise much above 6 miles in auroral regions, while ten times that generally is the polar lights' nearest, and usual, approach to the earth—nearest approach, not its outermost reach, for that is anything up to 600 miles, or more. Surprisingly, too, it is visible, and may be photographed, much farther out when its topmost portion is in sunshine than when it is in the shadow of the earth, of course when the sun is way below the horizon and the lower atmosphere therefore free from obscuring glare.

As already stated, the aurora is of various colors: silverwhite; pink; red—pale to deep and dense; green, of the mother- of-pearl or nacreous appearance; and yellowish. Often the same streak is red in its lower portion and greenish in its upper reaches. In shape, colour and motion the polar lights play varied and fascinating roles, all incident to electric discharges in the very high, rare and exceedingly ionized atmosphere. Most of this light long ago was traced to nitrogen, and recently, after many defeats, the rest of it to oxygen.

The more brilliant auroras commonly are accompanied by disturbances in the magnetism of the earth; also there are at these times marked electric currents in the earth, strong enough occasionally to interfere seriously with telegraphing. These facts make it all but certain that the auroral lights are caused by electric currents in the high atmosphere. This conclusion is further supported by the fact that the streamers of light follow the lines of the earth's magnetic force, just as an electric current would; and by the clinching fact that practically the only way to make oxygen and nitrogen give out light, the way they do at great heights, is by passing an electric current through them. That these luminous streamers parallel the earth's lines of magnetic force is shown by the fact that they all point to some place in the direction of the magnetic pole.

When there are several of them at the same time they seem to fan out from a common hub, or pivot, like streaks from the sun, only in this case the hub turns out to be the magnetic pole, and the fanning more apparent than real—a perspective effect like the seeming drawing closer together with increase of distance, of the parallel rails of a straight stretch of railway. This analogy is close, for the streaks are as straight as the rails, and the portions visible from any given place nearly parallel since their common hub usually is thousands of miles distant.

It also is known, from many years of observations, that auroral displays commonly are brightest and most extensive when a great sunspot squarely faces the earth, a fact that forces us to assume that the electric current in the extremely rarefied outer air that produces the aurora comes directly from the sun. And we believe, too, that this current is a stream of electrons—atomic negative electricity—or, occasion- ally, perhaps, positively charged atoms; for only such streams cause oxygen and nitrogen to glow as they do in the aurora.

How the heights of the auroras are measured deserves special and separate mention for its very cleverness and simplicity. Many auroras have clearly distinguishable, temporary markings or splotches, and often a fairly well defined and more enduring base. Evidently, therefore, it would be possible for observers at two stations, say 20 miles apart, in a known direction from each other, to record simultaneously the directions, horizontal and vertical, from their respective positions, of the same spot in an auroral streamer, and then with a little calculation of the easier sort compute its exact height above the earth.

In practice, however, the necessary observations consist of simultaneously photographing the aurora at the two stations against a background of stars (always visible through an aurora), and noting the exact time the exposures were made. Clearly the pictures of this same spot in the aurora will be at one place among the star images on the negative obtained at one of the stations, and at a different place on the negative made at the other station.

Then knowing these apparent positions of the spot in question among the stars, the exact time of the exposures, and the precise geographic locations at which they were made, one has only to turn to a suitable astronomical table to get the direction of the given spot from each place of observation, and quite as accurately as could be obtained with a theodolite (telescope provided with horizontal and vertical graduated circles). Finally, from these directions, and the direction from one station to the other, together with their distance apart, it is a simple and easy matter to compute the height of the given spot in the aurora.

This photographic method of determining auroral heights is longer than that by direct observations, but it obviates all the difficulty of obtaining simultaneous measurements, and all uncertainty as to whether the records of both stations apply to exactly the same point. It also enables one to study every distinguishable detail, and that, too, calmly, carefully and at any subsequent time. In short, though more tedious, it is much the better of the two methods.

LIGHTNING

The enchanting tracery of lightning over a distant summer-night's cloud; its blinding brilliance when alarmingly near; and the rolling rumble that goes with it, all, from ancient days, are recorded in history, related in legend and told in story. Its splendor, mystery and danger ever evoked admiration, reverence and awe. It may never itself have been an object of worship, as the sun has been and is, but momentarily, at least, it many a time compelled pious and prayerful thoughts—only that and nothing more.

It could not be gotten hold of like sticks and stones and thereby known as they were known, and so age followed age with no attempt whatever to account for lightning as a natural phenomenon. Nor, indeed, until comparatively recent times was there any possible way by which such an attempt could rationally be made. Sculpture, architecture and poetry had attained levels never yet surpassed, ages before any one had the slightest understanding of lightning; and all the world's great religions long had been preached and practiced, so much earlier, and more effectively, did we contemplate than investigate. The injunction of the ancient poet and philosopher was, "know thyself"; that of the modern scientist is, "know thou Nature." Both are productive of infinite good, but it was the latter, and it alone, that changed lightning for us from a terrifying, wilful act of an offended god to a normal and understandable natural phenomenon. For the former there were reasons aplenty; for the latter not a scintilla of evidence.

The first recorded hints from that great teacher, Nature, as to what lightning is were made, as her hints always are made, to those who *saw* what they saw—who comprehended the possible significance of what happened before their eyes. In this case Hawksbee published in 1705, and Wall in 1708, accounts of their experiments with frictional electricity, and how the phenomena thus obtained, though tiny and feeble, strangely resembled thunder and lightning.

From such experiments as these, made by many persons, it was surmised that thunder clouds are charged with electricity and that lightning is a manifestation of their discharge—surmised, but nothing more for as yet no experiment had been made that had any bearing on this problem, nor any even devised that would at all test the matter.

In 1749, however, Franklin, proposed a simple means "to determine," as he said, "the question, whether the clouds that contain lightning are electrified or not," but had no opportunity at that time, nor for some years thereafter, to convert his paper plans into physical equipment and subject it to the test of an actual experiment. But the plan itself had become well known, and on May 10, 1752, the Frenchman, Dalibard, proved by it that the clouds in which lightning occurs are electrified, and during the same summer Franklin confirmed this conclusion in a more convincing way —with his famous kite experiment.

There the matter rested, or practically just there, for decade after decade, until the scientist's curiosity induced, the industrialist's needs required, and new devices made possible, a fuller knowledge of lightning in its every phase, from genesis to dissipation. The search is on, keener than ever before, and more varied for there is no such thing as satisfying the investigator's curiosity—answer him one question and he asks a dozen other. Thus we learn.

If now we are going to talk about lightning, and we are, it will be well to follow the logical order in the famous receipt for cooking a rabbit pie, and first provide a cloud with the necessary electric charge. The fact that lightning occurs in a certain type of cloud and in no other, nor in that except when in an active state of precipitation, is so conspicuous that our veriest ancestral savage must have known it. But he did not know, and did not care, nor until recently did we know, however impatient our curiosity, how any cloud became electrically charged, nor what precipitation had to do with it.

One of the first leads, perhaps the first important direct lead, to the understanding of this phenomenon, as it now is more commonly explained, was found in 1892 by the German physicist, Philipp Lenard, when he observed that pure water, when splashed against a solid object, becomes positively charged and the near-by air negatively charged. This lead was followed to an open highway in 1908 by the English meteorologist, George C. Simpson, then in India, when he showed that the rupturing of drops of distilled water (ordinary tap water will not do, to which hangs a tale all its own) by a blast of air produces the same Lenard effect; the finer spray particles and the adjacent air being negatively charged, and the remaining larger parts of the drops positively charged. By George, be said—at least that is what any enthusiastic investigator would have said—that is how the thunderstorm cloud gets electrified. Rain water is distilled water, just the sort needed; check one.

Then, too, the uprush of air in a thunder cloud (cumulonimbus), as evidenced by the occurrence of hailstones, and testified by the aviator, is strong enough to keep raindrops suspended until they grow so big, about one fifth of an inch in diameter, that they are shaken to pieces by the wind that sweeps past them and carries the spray, at least, to still greater heights, an action that in the laboratory produces electrical separation; check two, and a long way, it would seem, on the road to a complete demonstration. But admitting, as we must, for the evidence is conclusive, that the air in a thunder cloud does rise with surprising velocity we naturally want to know why it does so, and at what level it will be breaking up raindrops at the maximum rate and thereby most rapidly building up the amount of electrical separation, or magnitudes of the electric as we commonly say, positive and negative. Well, it is just the gravity phenomenon of the heavier, or denser, pushing the lighter up, as water pushes up a cork that is let go beneath the surface.

Commonly this difference in the densities of adjacent masses of air, essential to convection, is effected in one or the other of two quite different ways. One is by local surface heating by which the air at the warmer place is brought to a higher temperature, and therefore is more expanded and lighter, volume for volume, than that round about. The other is by the importation over the surface of air colder than that surface. Increase in the amount of water vapour in the air also decreases its density, but generally this is of relatively little importance and therefore will not here be further considered.

In any case the lighter, or more exactly, the warmer air is pushed up by the colder, because it is the denser of the two. As this warmer air is forced up to higher levels it comes under less and less pressure to the extent of the weight of the air left below; and being under less pressure it expands, of course against whatever pressure may still be on it.

But this expansion against pressure means work, and work at the expense of the energy, or ability to do work, of the expanding air, that is, at the expense of its heat. It therefore cools with ascent, and often before it has gone a mile high its temperature has fallen to the dew point and condensation has begun. This condensation, in turn, sets free the heat that was used up in the first place in evaporating so much water as is now changed back from vapour to liquid, a surprising amount, approximately enough for every pound of cloud formed to heat six pounds of water from the freezing point to the boiling point. Yet, paradoxical as it may seem, this tremendous amount of heat does not increase the temperature of the air to which it is given, but it does prevent that air from cooling as rapidly with increase of height as it otherwise would, and thereby keeps the interior of the cloud all the time warmer than the air on the outside at the same level.

In this way a vigorous chimney-like up-draft is created within a cumulus cloud, maintained so long as there is an abundant inflow at the base of adequately humid air, and carried up to that height at which, incident to the resulting low temperature, but little vapour is left, too little, indeed, to furnish enough heat, through condensation, to push the convection to greater heights. The level at which the up-draft is strongest doubtless varies greatly from cloud to cloud, but from the fact that every foot of ascent develops an urge (though a progressively decreasing one) to further ascent, and also from the direct testimony of aviators, it seems that this level of swiftest up-rush normally is well above the cloud's midheight. The vigorous uprush of air in a thunderstorm cloud, therefore, is fully accounted for by the great amount of heat freed by, or resulting from, the abundant local condensation. Furthermore, we know too that the velocity of this ascent through much of the cloud's height is quite sufficient, 30 feet per second, or more, to carry along all water drops of smaller diameter than about one fifth of an inch.

Any larger drops that may form in this vigorously ascending air are blown to bits, just as such drops are in falling through still air on attaining a velocity of about 26 or 27 feet per second. The fragments broken off from the larger drops by the ascending current, are, of course, carried on to still greater heights. Hence the tendency is to accumulate a great amount of water at that level in the cloud where the ascending air just supports the largest drops that can remain whole. That is, the level at which the velocity of ascent is 25 to 30 feet per second.

This level appears usually to be above the midheight of the cloud, and since it necessarily is the place where electrical separation mainly occurs, and where the positive charge is chiefly concentrated, it follows that ordinarily most of the positive electricity in a thunderstorm cloud is in its upper portion. The negative electricity carried off on the spray from the ruptured drops canopies the top of the cumulus and, owing to the descent of the air all around it, drapes its sides. Hence a cloud charged in this way may be mainly positive in its lower portion, and negative in its upper portion, as previously was thought to be its usual state; or the other way around, positive above and negative below, as observation shows it generally is and as appears to be an entirely possible, if not, indeed, much the more likely result of electrical separation through the rupture of raindrops incident to vigorous convection.

As just implied there still is some uncertainty as to the exact process by which the towering thunderstorm cloud becomes electrified, but it is certain that such clouds do become tremendously charged, one portion with positive electricity and another with negative electricity. The two charges may be separated any distance, just as may artificial charges may far or near they never let up their pull on each other, partly through the nonconducting air directly, and indirectly, in part, through the conducting earth beneath. But close together with nearly all their mutual pull direct, or so far apart that the pull of each is essentially between itself and the charge it induces on the surface of the earth below, neither can be increased indefinitely any more than can a like charge on an object in the laboratory.

As the charge becomes greater and greater the electrical resistance finally breaks down—the crowded electric atoms burst their bounds much as gas molecules burst the walls of an over-filled balloon. As the charge builds up, or becomes more and more concentrated, any free electrons present (and there always are some) are moved with greater and greater speed by the force the increasing charge exerts upon them and in the general direction of the nearest large charge of the opposite sort. At first the impact of these electrons on the air particles (molecules) just bumps them a little but leaves them otherwise unaffected.

As the impelling charge grows greater, however, some of the electrons move with such speed that they no longer just rebound from the molecules they encounter, but disrupt them, at least to the extent of driving off one or more of their normal number of electrons. In this way the conductivity of the air into which the electrons are driven, or "shot," rapidly becomes greater and greater, the free electrons, incident to the increased bombardment, still more numerous and faster, and so on and on, every increase of either condition intensifying that of the other, until, in less time than it takes to bat an eve the selfboosting climaxes in a spasmodic rush of the whole charge, or much of it, at least, along a sinuous and often multi-branched narrow path which all but instantaneously becomes intensely luminous, and which we, therefore, call lightning, meaning, if we mean correctly, a flash of light induced by a natural electric discharge through the air.

The flash of light, as stated, is not coincident with the passage of the electricity, but is a consequence thereof. The discharge itself consists of extremely fast moving electric, or electrified, particles, electrons mainly, along a very crooked tube-like path. Along this tube, then, there are countless myriads of both free electrons and roving positive atoms—positive because partially stripped of their normal supply of electrons, with which they are neutral.

Instantly this separation of positive and negative charges is effected, they start recombining, each electric atom with its original partner, forming the same old neutral atom, or, more likely, with a different partner, and forming thereby a different but indistinguishable atom. They rush together violently, and a portion of the energy of their collision is given off in the form of light radiation. We conclude, therefore, from these laboratory determinations, based on many ingenious experiments, that the light along the path of an electric discharge is not owing immediately to the passage of the electricity, but to the quick automatic repair of the atoms it shattered along its course.

But whatever the process in detail of a lightning discharge within a cloud, or between a cloud and the earth, it is evident that only an exceptionally speedy device can adequately record it, for it is indeed "quick as lightning"—all over, a relativist might say, before it begins. Such a record is made photographically. The whole camera, for instance, is turned about a vertical axis at a known speed; or the light delivered to the camera lens by a rapidly rotating mirror; or two duplicate lenses equally distant from, and on opposite sides of, a common centre of rotation are rapidly spun about this centre in front of a stationary photographic plate; or the same, except that the lenses are stationary and the plate rotated; or by any equivalent device that will separate the images on the plate of closely sequent discharges, and show the time intervals between them when the rate of angular rotation of the moving element is given and focal lenght of the lens is known. It is fine sport, and several enthusiasts, playing the game most skillfully, have won at it a number of rich rewards.

In this way it has been learned that an electric discharge between a cloud and the earth commonly starts, like unto a small high speed projectile, from a negative lower portion of the cloud, over a devious course which it follows towards the earth at a uniform velocity around 125 miles per second, and along which so many wrecked atoms are left strewn as to render it electrically conducting. This lone pilot discharge is quickly followed and overtaken by rush after rush of electricity of the same sort traveling at the speed of roughly 12,000 miles per second, along the same path.

When the pilot is overtaken its speedy follower stops while it itself keeps on, but only to be overhauled again and again by a subsequent rush every time it has gotten ahead 33 feet, or so. At the same time fitful spurts of the opposite kind of electricity reach up from the earth until finally, that is, in the fiftieth of a second, perhaps, a conducting channel has been established from cloud to earth, whereupon there is a mighty rush of electricity up the prepared channel at the enormous velocity of roughly 25,000 miles per second (around the world in a second of time) and out any side branches that may have split off from it in the course of its development and extension from cloud to earth. This final, or return, rush affords much the brightest part of the entire lightning flash, and itself is brightest near the earth, and fades off with increase of distance along its course.

In many cases a lightning discharge is immediately followed by one or more others, as the flickering of its light indicates, generally at irregular intervals of a few hundredths of a second and along the same main course. These sequent discharges do not contain any of the intermediary pulsations, or gushes, only the initial pilot and the final return surge. Furthermore, only the first one or two of the sequent discharges, and commonly not even they, show any branches, and such as they do show are along the courses of the branches from the first discharge; and while the duration of any individual stroke is of the order of only one one-hundredth of a second, that of a sequence group may be anything up to an entire second, or even somewhat longer.

When the discharge is between cloud and cloud, or just from cloud to the space roundabout, it seems merely to diffuse along its course, or branched courses, into thin air until it, pilot and all, is too feeble to be perceptible. Neither, in this case, is there any return stroke, so far as we know—certainly nothing like the brilliant return sent up from the earth.

The lightning above described generally is called *streak lightning*, for the obvious reason that it occurs in streaks; not straight, nor zigzag with sharp angles and straight links, as artists used to represent it, but crooked with rounded or gradual turns. It also frequently is called *forked lightning*, in allusion, of course, to its appearance when divided into two or more branches. Another common name for it is *chain lightning*, having reference to its many apparent lines or segments. Sometimes a streak of lightning has seemed to lengthen at a perceptible speed instead of appearing to be all at once with no part first. Such real, or seeming, discharges have been called *rocket lightning*. It has not often been reported, and indeed may be merely an illusion.

On rather rare occasions a lightning streak persists for half a second, or longer like a suspended glowing wire, and then on fading away seems to divide into many bright and dark segments. The bright parts evidently are those viewed more or less end-on by the observer, since to him the light from the whole of each such portion appears to come from nearly the same spot, which spot therefore to him is relatively bright; while the dark (less bright) segments, on the other hand, are the portions of the streak which are rather across the observer's vision—segments whose light reaches the observer diffusely from a line and not concentratedly as from a point. Naturally, therefore, such streaks might be called, and they are called, *beaded lightning* and *pearl lightning*.

A far more frequent apparent form of lightning is the beautiful glow, momentary or briefly flickering, that often is seen of a summer's night in a distant cumulus cloud. This is not a special and distinct form of lightning as often has been assumed, and to which the names *sheet lightning* and *heat lightning* have been given, but just the diffusely reflected light from a streak invisible to the observer.

All the above forms of lightning, then, reduce to but one, streak lightning, and there remains just a single additional form to consider, namely, *ball liqhtning*—a humbug. Most cases of ball lightning are just after images, due to persistency of vision, from an instantaneous brilliant flash caught by the eye, similar to the after image one has of the sun following a momentary unprotected glance at it. One man reports find- ing this explanation for himself years after seeing ball lightning pass so close to him that he "felt the heat from it," but to which a cow he was milking at the time gave no attention. Of course not, as he himself now says, since the whole thing was in his own eye, and not an objective reality which the cow could experience. Another thing often called ball lightning is just a brush discharge, common on high peaks especially, during thunderstorm weather. Several other phenomena also have occasionally been reported as ball lightning, but not one of them has stood the test of careful examination.

All lightning then appears to be streak lightning, and every one wants to know how long the streak is, or can be. If we decide that brush discharges and tiny sparks, wherever they occur and however induced, are not lightning, or that even if they are they are not included under this query, but only streaks that may be seen at a considerable distance, then the answer is: Any length from perhaps 100 yards up to at least 12 miles, in the case of a meandering streak, stretched out to its full length.

It is natural, too, to want to know how hot the air is along the lightning's path, but this question cannot be answered in terms of definite figures on a thermometer scale. We can only say that some of the things that happen incident to the discharge, such as the explosion-like expansion that causes thunder, are just what one would expect from a suddenly produced high temperature. In short its effective temperature is very great.

Well, if the temperature is exceedingly high along a streak of lightning, doesn't the air there get scorched? Yes, in a sense, and many people say they can smell the scorched, or burnt, aiı and that always the odor is like that of burnt sulphur.

The fact is that the lightning disrupts many of the oxygen, nitrogen and water vapour molecules along its course and that in their resultant combination with whatever is present several substances are formed that were not there before, especially combinations of nitrogen with hydrogen and with oxygen. In the presence of water vapour, always in the air, the first of these goes over into ammonia, and the second into nitrous and nitric acids. All these things, ammonia, the oxides of nitrogen and the nitrogen acids, are pungent, and therefore are mistaken by most people who are not familiar with them for the highly pungent fumes of burning sulphur with which they, like the rest of us, are familiar.

It is interesting to note here that this ammonia promotes plant growth, and that both the nitrogen acids, on reaching the ground, produce compounds with substances in the soil which are excellent plant food, from which facts one correctly infers that thunderstorms above enrich the soil beneath. However, thunderstorm frequency and soil fertility do not in general run, even approximately, parallel to each other since there are many other factors that contribute to the final result. Lightning only makes the soil richer than it otherwise would be.

Lightning however, can do more things than furnish rich food for plants and make gloriously beautiful an otherwise dark and somber cloud. It can rend, crush, explode, fuse, and fire objects that it actually his-in the twinkling of an eye tear to slivers a giant tree that for centuries had but grown bigger and stronger with every stress and strain wind and rain and storm did bring. Naturally we wonder how such sudden and complete destruction is effected. The answer is: By an explosion.

When the current only follows the grain of the wood along a narrow strip just under the bark, the damage often is rather slight—just the bark over that strip ripped off and the solid wood beneath only slightly slivered—but when a heavy discharge follows the centre of the tree, as it sometimes does, the result may be the collapse of the entire trunk and the scattering of its parts hundreds of feet away in every direction. Here along the course of the discharge whatever poorly conducting substances are actually encountered are, in large measure, not only volatilized but also ionized. Hence the volume of these substances, sometimes along the very heart of a tree, is made many fold greater with all the suddenness and violence, and therefore, destructive effects, of exploding dynamite, that is, dynamite suddenly reverting to its constituents in the gaseous form, and requiring forthwith a greatly larger volume.

If the object encountered is a conductor, anyone of three things may happen, depending on its size and nature. A fine wire, for instance, is completely volatilized. It may even be blown as a copper gas, say, through an insulating cloth jacket without scorch or other injury, leaving an apparent insulated wire that is all hollow insulation and nothing more, the wire having "spurlos" (without trace) vanished. A wire of medium size, however, may just fuse, or only get quite hot, and, in either case, scorch its insulating jacket, if so covered. Finally, a metal rod is likely to be only warmed more or less, except at the points of entrance and exit where there is certain to be some fusion and volatilization.

On rare occasions lightning produces an effect just the reverse of an explosion—a collapse. This is conspicuous when a metal (copper, say) tube is struck with a current just sufficient to render its wall plastic, in which case the tube is crushed in as if by an explosion all around it. Evidently there is some reactional push in as a result of the sudden expansion of the abruptly heated surrounding air, but the air within the tube will, for the same reason, push out quite as hard, if not harder. Hence the collapse of the tube must be owing to some other cause, and that cause has been found to be a magnetic squeeze between all portions of the wall that are simultaneously carrying a current of the same sort in the same direction.

This conclusion follows from the fact that a wire carrying a current is, so long as the current lasts, surrounded by a magnetic field of such nature that two such wires parallel to each other draw as close together as possible, and that a bundle of such wires pull themselves into a compact strand. If now we regard the wall of our tube as consisting of a bundle of wires, all simultaneously carrying currents of the same sort in the same direction, it will be clear that there must be a radial compression on it tending to crush the wall into an axial rod, and that it will be so crushed as soon as rendered quite plastic, as sometimes happens incident to heating by the strong current, 20,000 amperes, or more of a lightning discharge.

Since lightning can smash a huge tree to bits with the violence and suddenness of a terrific explosion, and do many other things that require herculean force to perform, it is not surprising that many practical people have insisted that lightning no longer be permitted to just look pretty and make a noise, but, like Niagara, harnessed to some prosaic machine and made to do menial work. Obviously, if one should undertake to obtain financial backing for such an undertaking as this, he would do well to have some idea as to how much energy there is in the average lightning discharge.

From the strength of the current, say 20,000 amperes, the electric push, or voltage, between cloud and earth, around 25,000 volts per inch, and the total duration of the discharge, say one one-thousandth of a second, we find that if the streak is half a mile long, the energy used up is the equivalent of about 2,500 horsepower continuously for 24 hours, and as there are, over the earth as a whole, some 44,000 flashes of lightning per day, we may divide our result by 10, to include minor discharges and provide a large margin of safety, and still have for the world's average total lightning energy the equivalent of fully 1,100,000,000 horsepower, continuously, day and night, year in and year out.

At the low rate of one cent per kilowatt hour this would come, in round numbers, to $200,000,000 per day. Easy calculation, this, and tempting to the man of means; but the electrical engineer will find his part of the job something like building a bridge to the moon. Nevertheless, the facts in the case are of the kind to incite many a pleasing fancy, and to make scientifically plausible endless alleged happenings more amazing and spectacular than ever Aladdin could effect, rub his lamp and his ring as he might.

It was stated above that there really is only one sort of lightning, bright streak lightning, and yet we frequently are asked to explain dark, or black, lightning, and if we venture to insist that there isn't any such thing we are promptly shown a picture that is presumed to put us in the awkward dilemma of admitting either inexcusable ignorance or wilful prevarication.

We don't have to admit either alternative, however, for the dark streaks on the picture where one would expect white ones, are the result of a curious photographic reversal, as it is called. When an exposure to one lightning discharge is followed by diffuse light from others, such as normally would "fog" the plate, the fainter initial streaks are reversed on the negative and therefore print black, while the heavier portions of the main streak image are reversed only along their borders and therefore print as white lines with dark edges. No, lightning itself is not dark, but dazzlingly bright, even though circumstances at the time may cause its picture to show up dark.

The above various effects and phenomena of lightning might seem to pretty thoroughly cover the field, but it has also a noisy by-product, thunder, that likewise clamors for a hearing. As previously explained, myriads of molecules all along a lightning discharge are torn apart into their constituent atoms, and vast numbers of the atoms stripped each of one or more of its electrons, everyone of which parts, atom, stripped atom and electron requires as much space as would the original whole, or molecule, at the same temperature and pressure-and the effective temperature is now very high.

Hence from end to end of the lightning path there is an abrupt and violent explosion, so sudden that the surrounding air cannot just flow away gently, but only crowd and compress without quarter. Each crowded sheath around the path taken by the discharge, passes the push, with its consequent crowding and compression, on to the next, the next and the next until the volume affected becomes so large that the compression is negligible and the noise (thunder) inaudible. The distance at which this occurs, that is, the distance to which thunder can be heard, necessarily varies greatly with the course of the stroke, whether cloud to earth (loudest), cloud to cloud or open space; volume of the discharge; height of streak above sea level (density of the air in which it occurs); direction and strength of the local winds, and anything else that can affect the carry of sound. However, it seldom is heard more than ten to twelve miles from its source, which, as we know, is not a fourth the distance cannons often are heard.

This difference is owing mainly to the fact that all the cannon's noise comes from one place, the mouth of the cannon, while that of the lightning is from a crooked line a mile long or more, only the nearest portion of which gives the first, and, normally, loudest crash. That is, we have the benefit of the cannon's total effort, but of only a small fraction, at any one instant, of that of the lightning.

Another reason why a cannon often is heard surprisingly farther than thunder is the fact that the cannon may be fired when the air is still, and even in the early morning when the lower air is cool and therefore in its best condition for carrying sound, while thunder occurs only when the air is turbulent, a condition that causes it to scatter and diffuse sound much as crinkled glass wastes light.

The cannon, however, gives but a single boom, while thunder keeps on rumbling and roaring for many seconds, even nearly an entire minute in some cases, owing to the fact that the distance to the farthest portion of the lightning streak often is several miles (it takes sound five seconds to travel a mile) greater than that to its closest point. Clearly, too, when any appreciable portion of the streak is throughout at about the same distance from one he hears the noise from all of that portion simultaneously, and therefore as a loud boom. This is the main cause of the rumbling of thunder. In favorable localities there is more or less echoing of thunder by cliffs and other obstacles, but this effect commonly is of small importance. Especially is the echoing of thunder by clouds, so often alleged, practically negligible, as the mathematical analysis of the subject conclusively proves, and as anyone can easily verify by yelling at a dense fog bank and listening for the echo, an echo that never comes, unless from a house, cliff, or other solid object, in the fog.

Chapter 9

Earthquake

The sudden movement of the Earth caused by the abrupt release of accumulated strain along a fault in the interior. The released energy passes through the Earth as seismic waves (low-frequency sound waves), which cause the shaking. Seismic waves continue to travel through the Earth after the fault motion has stopped. Recordings of earthquakes, called seismograms, illustrate that such motion is recorded all over the Earth for hours, and even days, after an earthquake.

Earthquakes are not distributed randomly over the globe but tend to occur in narrow, continuous belts of activity. Approximately 90% of all earthquakes occur in these belts, which define the boundaries of the Earth's plates. The plates are in continuous motion with respect to one another at rates on the order of centimeters per year; this plate motion is responsible for most geological activity.

Plate motion occurs because the outer cold, hard skin of the Earth, the lithosphere, overlies a hotter, soft layer known as the asthenosphere. Heat from decay of radioactive minerals in the Earth's interior sets the asthenosphere into thermal convection. This convection has broken the lithosphere into plates which move about in response to the convective motion. As the plates move past each other, little of the motion at their boundaries occurs by continuous slippage; most of the motion occurs in a series of rapid jerks. Each jerk is an earthquake. This happens because, under the pressure and temperature conditions of the shallow part of the Earth's lithosphere, the frictional sliding of rock exhibits a property known as stick-slip, in which frictional sliding occurs in a series of jerky movements, interspersed with periods of no motion—or sticking. In the geologic time frame, then, the lithospheric plates chatter at their boundaries, and at any one place the time between chatters may be hundreds of years.

The periods between major earthquakes is thus one during which strain slowly builds up near the plate boundary in response to the continuous movement of the plates. The strain is ultimately released by an earthquake when the frictional strength of the plate boundary is exceeded.

Most great earthquakes occur on the boundaries between lithospheric plates and arise directly from the motions between the plates. These may be called plate boundary earthquakes. There are many earthquakes, sometimes of substantial size, that cannot be related so simply to the movements of the plates. At many plate boundaries, earthquakes occur over a broad zone—often several hundred miles wide—adjacent to the plate boundary. These earthquakes, which may be called plate boundary-related earthquakes, are secondarily caused by the stresses set up at the plate boundary. Some earthquakes also occur, although infrequently, within plates. These earthquakes, which are not related to plate boundaries, are called intraplate earthquakes. The immediate cause of intraplate earthquakes is not understood.

In addition to the tectonic types of earthquakes described above, some earthquakes are directly associated with volcanic activity. These volcanic earthquakes result from the motion of undergound magma that leads to volcanic eruptions.

Earthquakes often occur in well-defined sequences in time. Tectonic earthquakes are often preceded, by a few days to weeks, by several smaller shocks (foreshocks), and are nearly always followed by large numbers of aftershocks. Foreshocks and aftershocks are usually much smaller than the main shock. Volcanic earthquakes often occur in flurries of activity, with no discernible main shock. This type of sequence is called a swarm.

Earthquakes range enormously in size, from tremors in which slippage of a few tenths of an inch occurs on a few feet of fault, to the greatest events, which may involve a rupture many hundreds of miles long, with tens of feet of slip.

The size of an earthquake is given by its moment: average slip times the fault area that slipped times the elastic constant of the Earth. The units of seismic moment are dyne-centimeters. An older measure of earthquake size is magnitude, which is proportional to the logarithm of moment. Magnitude 2.0 is about the smallest tremor that can be felt. Most destructive earthquakes are greater than magnitude 6; the largest shock known was the 1960 Chile earthquake, with a moment of 10^{30} dyne-centimeters (10^{23} newton-meters) or magnitude 9.5. It involved a fault 600 mi (1000 km) long slipping 30 ft (10 m).

The intensity of an earthquake is a measure of the severity of shaking and its attendant damage at a point on the surface of the Earth. The same earthquake may therefore have different intensities at different places. The intensity usually decreases away from the epicenter (the point on the surface directly above the onset of the earthquake), but its value depends on many factors and generally increases with moment. Intensity is usually higher in areas with thick alluvial cover or landfill than in areas of shallow soil or bare rock. Poor building construction leads to high intensity ratings because the damage to structures is high. Intensity is therefore more a measure of the earthquake's effect on humans than an innate property of the earthquake.

An earthquake is the result of a sudden release of energy in the Earth's crust that creates seismic waves. Earthquakes are recorded with a seismometer, also known as a seismograph. The moment magnitude of an earthquake is conventionally reported, or the related and mostly obsolete Richter magnitude, with magnitude 3 or lower earthquakes being mostly imperceptible and magnitude 7 causing serious damage over large areas. Intensity of shaking is measured on the modified Mercalli scale.

At the Earth's surface, earthquakes manifest themselves by shaking and sometimes displacing the ground. When a large earthquake epicenter is located offshore, the seabed sometimes suffers sufficient displacement to cause a tsunami. The shaking in earthquakes can also trigger landslides and occasionally volcanic activity.

In its most generic sense, the word *earthquake* is used to describe any seismic event—whether a natural phenomenon or an event caused by humans—that generates seismic waves. Earthquakes are caused mostly by rupture of geological faults, but also by volcanic activity, landslides, mine blasts, and nuclear experiments. An earthquake's point of initial rupture is called its focus or hypocenter. The term epicenter refers to the point at ground level directly above this.

NATURALLY OCCURRING EARTHQUAKES

Tectonic earthquakes will occur anywhere within the earth where there is sufficient stored elastic strain energy to drive fracture propagation along a fault plane. In the case of transform or convergent type plate boundaries, which form the largest fault surfaces on earth, they will move past each other smoothly and aseismically only if there are no irregularities or asperities along the boundary that increase the frictional resistance. Most boundaries do have such asperities and this leads to a form of stick-slip behaviour.

Once the boundary has locked, continued relative motion between the plates leads to increasing stress and therefore, stored strain energy in the volume around the fault surface. This continues until the stress has risen sufficiently to break through the asperity, suddenly allowing sliding over the locked portion of the fault, releasing the stored energy. This energy is released as a combination of radiated elastic strain seismic waves, frictional heating of the fault surface, and cracking of the rock, thus causing an earthquake.

This process of gradual build-up of strain and stress punctuated by occasional sudden earthquake failure is referred to as the Elastic-rebound theory. It is estimated that only 10 per cent or less of an earthquake's total energy is radiated as seismic energy. Most of the earthquake's energy is used to power the earthquake fracture growth or is converted into heat generated by friction. Therefore, earthquakes lower the Earth's available elastic potential energy and raise its temperature, though these changes are negligible compared to the conductive and convective flow of heat out from the Earth's deep interior.

CAUSES OF EARTHQUAKE

An Earthquake is a sudden tremor or movement of the earth's crust, which originates naturally at or below the surface. The word natural is important here, since it excludes shock waves caused by French nuclear tests, man made explosions and landslides caused by building work.

There are two main causes of earthquakes:

- Firstly, they can be linked to explosive volcanic eruptions; they are in fact very common in areas of volcanic activity where they either proceed or accompany eruptions.
- Secondly, they can be triggered by Tectonic activity associated with plate margins and faults. The majority of earthquakes world wide are of this type.

An earthquake can be likened to the effect observed when a stone is thrown into water. After the stone hits the water a series of concentric waves will move outwards from the centre. The same events occur in an earthquake. There is a sudden movement within the crust or mantle, and concentric shock waves move out from that point. Geologists and Geographers call the origin of the earthquake the focus. Since this is often deep below the surface and difficult to map, the location of the earthquake is often referred to as the point on the Earth surface directly above the focus. This point is called the epicentre.

The strength, or magnitude, of the shockwaves determines the extent of the damage caused. Two main scales exist for defining the strength, the Mercalli Scale and the Richter Scale.

Earthquakes are three dimensional events, the waves move outwards from the focus, but can travel in both the horizontal and vertical plains. This produces three different types of waves which have their own distinct characteristics and can only move through certain layers within the Earth. Lets take a look at these three forms of shock waves.

TYPES OF SHOCKWAVES

P-Waves

Primary Waves (P-Waves) are identical in character to sound waves. They are high frequency, short-wavelength, longitudinal waves which can pass through both solids and liquids. The ground is forced to move forwards and backwards as it is compressed and decompressed. This produces relatively small displacements of the ground.

P Waves can be reflected and refracted, and under certain circumstances can change into S-Waves.

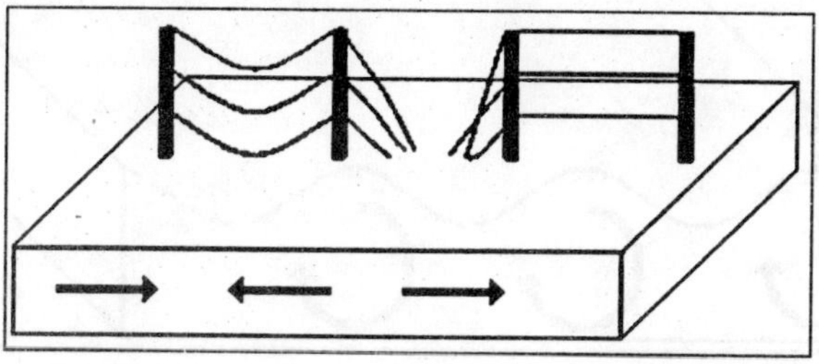

Fig. Particles are Compressed and Expanded in the Wave's Direction.

S-Waves

Secondary Waves (S-Waves) travel more slowly than P-Waves and arrive at any given point *after* the P-Waves. Lik e P-Waves they are high frequency, short-wavelength waves, but instead of being longitudinal they are transverse. They move in all directions away from their source, at speeds which depend upon the density of the rocks through which they are moving. They cannot move through liquids. On the surface of the Earth, S-Waves are responsible for the sideways displacement of walls and fences, leaving them 'S' shaped.

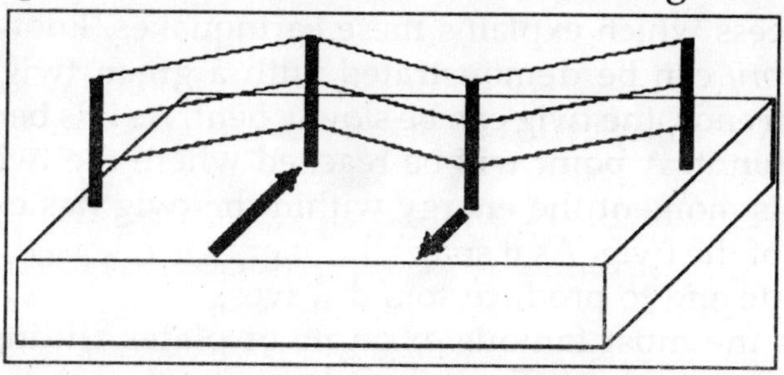

Fig. S-Waves Move Particles at 90° to the Wave's Direction.

L-Waves

Surface Waves (L-Waves) are low frequency transverse vibrations with a long wavelength. They are created close to the epicentre and can only travel through the outer part of the crust. They are responsible for the majority of the building damage caused by earthquakes. This is because L Waves have a motion similar to that of waves in the sea. The ground is made to move in a circular motion, causing it to rise and fall as visible waves move across the ground. Together with secondary effects such as landslides, fires and tsunami these waves account for the loss of approximately 10,000 lives and over $100 million per year.

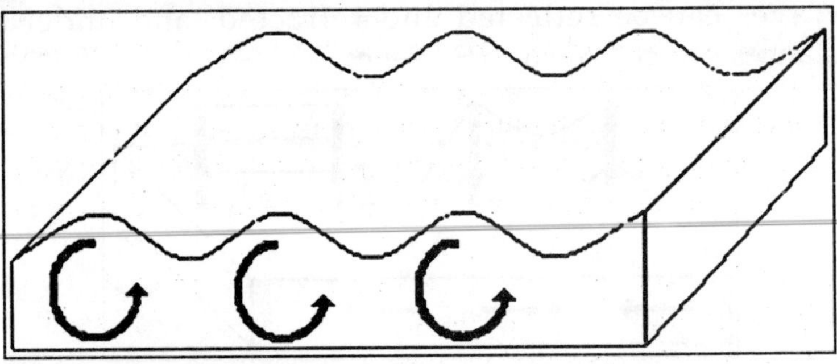

Fig. L-Waves Move Particles in a Circular Path.

Tectonic Earthquakes

Tectonic earthquakes are triggered when the crust becomes subjected to strain, and eventually moves. The theory of plate tectonics explains how the crust of the Earth is made of several plates, large areas of crust which float on the Mantle. Since these plates are free to slowly move, they can either drift towards each other, away from each other or slide past each other. Many of the earthquakes which we feel are located in the areas where plates collide or try to slide past each other.

The process which explains these earthquakes, known as *Elastic Rebound Theory* can be demonstrated with a green twig or branch. Holding both ends, the twig can be slowly bent. As it is bent, energy is built up within it. A point will be reached where the twig suddenly snaps. At this moment the energy within the twig has exceeded the *Elastic Limit* of the twig. As it snaps the energy is released, causing the twig to vibrate and to produce sound waves.

Perhaps the most famous example of plates sliding past each other is the San Andreas Fault in California. Here, two plates, the Pacific Plate and the North American Plate, are both moving in a roughly northwesterly direction, but one is moving faster than the other. The San Francisco area is subjected to hundreds of small earthquakes every year as the two plates grind against each other. Occasionally, as in 1989, a much larger movement occurs, triggering a far more violent 'quake'.

Major earthquakes are sometimes preceded by a period of changed activity. This might take the form of more frequent minor shocks as the rocks begin to move,called *foreshocks,* or a period of less frequent shocks as the two rock masses temporarily 'stick' and become locked together.

Detailed surveys in San Francisco have shown that railway lines, fences and other longitudinal features very slowly become deformed as the pressure builds up in the rocks, then become noticeably offset when a movement occurs along the fault. Following the main shock, there may be further movements, called *aftershocks,* which occur as the rock masses 'settle down' in their new positions. Such aftershocks cause problems for rescue services, bringing down buildings already weakened by the main earthquake.

Volcanic Earthquakes

Volcanic earthquakes are far less common than Tectonic ones. They are triggered by the explosive eruption of a volcano. Given that not all volcanoes are prone to violent eruption, and that most are 'quiet' for the majority of the time, it is not surprising to find that they are comparatively rare.

When a volcano explodes, it is likely that the associated earthquake effects will be confined to an area 10 to 20 miles around its base, where as a tectonic earthquake may be felt around the globe.

The volcanoes which are most likely to explode violently are those which produce *acidic* lava. Acidic lava cools and sets very quickly upon contact with the air. This tends to chock the volcanic vent and block the further escape of pressure. For example, in the case of Mt Pelee, the lava solidified before it could flow down the sides of the volcano. Instead it formed a spine of solid rock within the volcano vent. The only way in which such a blockage can be removed is by the build up of pressure to the point at which the blockage is literally exploded out of the way. In reality, the weakest part of the volcano will be the part which gives way, sometimes leading to a sideways explosion as in the Mt St.Helens eruption.

When extraordinary levels of pressure develop, the resultant explosion can be devastating, producing an earthquake of considerable magnitude. When Krakatoa (Indonesia, between Java and Sumatra) exploded in 1883, the explosion was heard over 5000 km away in Australia. The shockwaves produced a series of tsunami (large sea waves), one of which was over 36m high; that's the same as four, two story houses stacked on top of each other. These swept over the coastal areas of Java and Sumatra killing over 36,000 people.

By contrast, volcanoes producing free flowing *basic* lava rarely cause earthquakes. The lava flows freely out of the vent and down the sides of the volcano, releasing pressure evenly and constantly. Since pressure doesn't build up, violent explosions do not occur.

FAULT

In geology, a fault or fault line is a planar fracture in rock in which the rock on one side of the fracture has moved with respect to the rock on the other side. Large faults within the Earth's crust are the result of differential or shear motion and active fault zones are the causal locations of most earthquakes. Earthquakes are caused by energy release during rapid slippage along a fault. A fault that runs along the boundary between two tectonic plates is called a transform fault.

Since faults do not usually consist of a single, clean fracture, the term *fault zone* is used when referring to the zone of complex deformation that is associated with the fault plane. The two sides of a non-vertical fault are called the *hanging wall* and *footwall*. By definition, the hanging wall occurs above the fault and the footwall occurs below the fault. This terminology comes from mining. When working a tabular ore body the miner stood with the footwall under his feet and with the hanging wall hanging above him.

The creation and behaviour of faults, in both an individual small fault and within the greater fault zones which define the tectonic plates, is controlled by the relative motion of rocks on either side of the fault surface.

Because of friction and the rigidity of the rock, the rocks cannot simply glide or flow past each other. Rather, stress builds up in rocks and when it reaches a level that exceeds the strain threshold, the accumulated potential energy is released as strain, which is focused into a plane along which relative motion is accommodated — the fault.

Strain is both accumulative and instantaneous depending on the rheology of the rock; the ductile lower crust and mantle accumulates deformation gradually via shearing whereas the brittle upper crust reacts by fracture, or instantaneous stress release to cause motion along the fault. A fault in ductile rocks can also release instantaneously when the strain rate is too great. The energy released by instantaneous strain release is the cause of earthquakes, a common phenomenon along transform boundaries.

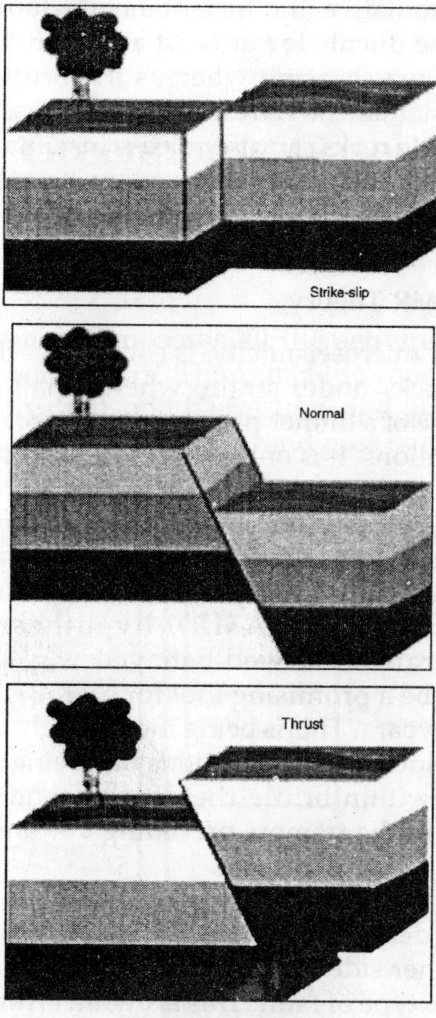

Fig. Fault Types

Mechanics

The creation and behaviour of faults, in both an individual small fault and within the greater fault zones which define the tectonic plates, is controlled by the relative motion of rocks on either side of the fault surface.

Because of friction and the rigidity of the rock, the rocks cannot simply glide or flow past each other. Rather, stress builds up in rocks and when it reaches a level that exceeds the strain threshold, the accumulated potential energy is released as strain, which is focused into a plane along which relative motion is accommodated — the fault.

Strain is both accumulative and instantaneous depending on the rheology of the rock; the ductile lower crust and mantle accumulates deformation gradually via shearing whereas the brittle upper crust reacts by fracture, or instantaneous stress release to cause motion along the fault. A fault in ductile rocks can also release instantaneously when the strain rate is too great. The energy released by instantaneous strain release is the cause of earthquakes, a common phenomenon along transform boundaries.

Microfracturing and AMR Theory

Microfracturing, or microseismicity, is sometimes thought of as a symptom caused by rocks under strain, where small-scale failures, perhaps on areas the size of a dinner plate or a small area, release stress under high strain conditions. It is only when sufficient microfractures link up into a large slip surface that a large seismic event or earthquake can occur. According to this theory, after a large earthquake, the majority of the stress is released and the frequency of microfracturing is exponentially lower. A related theory, accelerating moment release (AMR), hypothesizes that the seismicity rate accelerates in a well-behaved way prior to large earthquakes, and may be a promising tool for earthquake prediction on the scale of days to years. This is being increasingly used to predict rock failures within mines and applications are being attempted for the portions of faults within brittle rheological conditions. Similar behaviour is observed in the tremors preceding volcanic eruptions.

Slip, Heave, Throw

T he sense of slip is defined by the relative movements of geological features present on either side of the fault plane and is a vector. The sense of slip defines the type of fault. This is distinct from the throw of the fault, which is the vertical offset. Heave is the measured horizontal offset of the fault. The vector of slip can be qualitatively measured by fault bend folding, i.e. drag folding of strata on either side of the fault; the direction and magnitude of heave and throw can be measured only by finding common intersection points on either side of the fault. In practise it is usually only possible to find the slip direction of faults, and an approximation of the heave and throw vector.

Fault Types

Faults can be categorized into three groups based on the sense of slip. A fault where the main sense of movement (or slip) on the fault plane is vertical is known as a dip-slip fault. Where the main sense of slip is horizontal the fault is known as a transcurrent or strike-slip fault. Oblique-slip faults have significant components of both strike and dip slip.

For all naming distinctions, it is the orientation of the net dip and sense of slip of the fault which must be considered, not the present-day orientation, which may have been altered by local or regional folding or tilting.

Dip-Slip Faults

Dip-slip faults can be again classified into the types "reverse" and "normal". A normal fault occurs when the crust is extended. Alternatively such a fault can be called an extensional fault. The hanging wall moves downward, relative to the footwall. A downthrown block between two normal faults dipping towards each other is called a graben. An upthrown block between two normal faults dipping away from each other is called a horst. Low-angle normal faults with regional tectonic significance may be designated detachment faults.

A reverse fault is the opposite of a normal fault — the hanging wall moves up relative to the footwall. Reverse faults are indicative of shortening of the crust. The dip of a reverse fault is relatively steep, greater than 45°.

A thrust fault has the same sense of motion as a reverse fault, but with the dip of the fault plane at less than 45°. Thrust faults typically form ramps, flats and fault-bend (hanging wall and foot wall) folds. Thrust faults are responsible for forming nappes and klippen in the large thrust belts.

The fault plane is the plane that represents the fracture surface of a fault.

Flat segments of thrust fault planes are known as *flats,* and inclined sections of the thrust are known as *ramps*. Typically thrust faults move *within* formations by forming flats, and climb up section with ramps.

Fault-bend folds are formed by movement of the hangingwall over a non-planar fault surface and are found associated with both extensional and thrust faults.

Faults may be reactivated at a later time with the movement in the opposite direction to the original movement (fault inversion).

A normal fault may therefore become a reverse fault and vice versa.

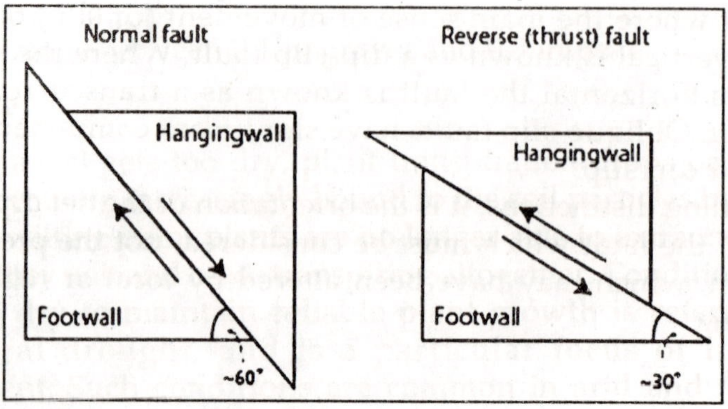

Fig. Cross-Sectional Illustration of Normal and Reverse dip-slip Faults.

Strike-Slip Faults

The fault surface is usually near vertical and the footwall moves either left or right or laterally with very little vertical motion. Strike-slip faults with left-lateral motion are also known as *sinistral* faults. Those with right-lateral motion are also known as *dextral* faults. A special class of strike-slip faults is the transform faults which are a plate tectonics feature related to spreading centers such as mid-ocean ridges.

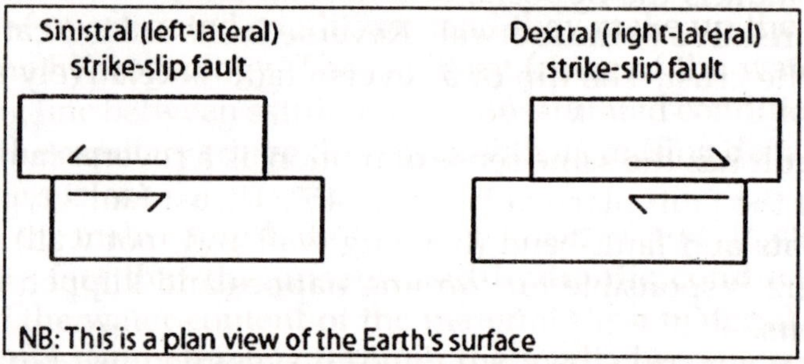

Fig. Schematic Illustration of the Two Strike-slip Fault Types.

Oblique-Slip Faults

A fault which has a component of dip-slip and a component of strike-slip is termed an oblique-slip fault. Nearly all faults will have some component of both dip-slip and strike-slip, so defining a fault as oblique requires both dip and strike components to be measurable and significant. Some oblique faults occur within transtensional and transpressional regimes, others occur where the direction of extension or shortening changes during the deformation but the earlier formed faults remain active.

HADE ; Hade is defined as the complement of dip, it is the angle between the fault plane and a vertical plane that strikes parallel to the fault.

Fault Rock

All faults have a measurable thickness, made up of deformed rock that is characteristic of the level in the crust where the faulting happened, the rock types affected by the fault and the presence and nature of any mineralising fluids. Fault rocks are classified by their textures and the implied mechanism of deformation. A fault that passes through different levels of the lithosphere will have many different types of fault rock developed along its surface. Continued dip-slip displacement will tend to juxtapose fault rocks characteristic of different crustal levels, with varying degrees of overprinting. This effect is particularly clear in the case of detachment faults and major thrust faults.

The main types of fault rock are:

- Cataclasite - A fault rock which is cohesive with a poorly developed or absent planar fabric, or which is incohesive, characterised by generally angular clasts and rock fragments in a finer-grained matrix of similar composition.
- Mylonite - A fault rock which is cohesive and characterized by a well developed planar fabric resulting from tectonic reduction of grain size, and commonly containing rounded porphyroclasts and rock fragments of similar composition to minerals in the matrix
- Tectonic or Fault Breccia - A medium- to coarse-grained cataclasite containing >30% visible fragments.
- Fault Gouge - An incohesive, clay-rich fine- to ultrafine-grained cataclasite, which may possess a planar fabric and containing <30% visible fragments. Rock clasts may be present
- Pseudotachylite - Ultrafine-grained vitreous-looking material, usually black and flinty in appearance, occurring as thin planar veins, injection veins or as a matrix to pseudoconglomerates or breccias, which infills dilation fractures in the host rock.
- Clay smear clay-rich fault gouge formed in sedimentary sequences containing clay-rich layers which are strongly deformed and sheared into the fault gouge.

AFTERSHOCK

An aftershock is an earthquake that occurs after a previous earthquake (the main shock). An aftershock is in the same region of the main shock but is always of smaller magnitude strength. If an aftershock is larger than the main shock, the aftershock is redesignated as the main shock and the original main shock is redesignated as a foreshock. Aftershocks are smaller earthquakes formed as the crust around the displaced fault plane adjusts to the effects of the main shock.

Distribution of Aftershocks

Most aftershocks are located over the full area of fault rupture and either occur along the fault plane itself or along other faults within the volume affected by the strain associated with the main shock. Typically, aftershocks are found up to a distance equal to the rupture length away from the fault plane. The pattern of aftershocks helps confirm the size of area that slipped during the main shock. In the case of the 2004 Indian Ocean earthquake and the 2008 Sichuan earthquake the aftershock distribution shows that in both cases that the epicenter (where the rupture initiated) lies to one end of the final area of slip, implying strongly asymmetric rupture propagation.

Aftershock Size and Frequency with Time

Aftershocks tend to obey a number of empirical laws concerning magnitude and frequency.

Omori's Law

Aftershocks occur with a pattern that follows Omori's law. Omori's law, or more correctly the modified Omori's law, is an empirical relation for the temporal decay of aftershock rates. In 1894, Omori published his work on the aftershocks of earthquakes, in which he stated that aftershock frequency decreases by roughly the reciprocal of time after the main shock.

$$n(t)=\frac{K}{c+t}$$

where:

- *n(t)* is the rate of earthquakes measured in a certain time *t* after the main shock,
- *K* is the amplitude, and
- *c* is the "time offset" parameter.

The modified version of Omori's law, now commonly used, was proposed by Utsu in 1961.

$$n(t)=\frac{K}{(c+t)^{P}}$$

where

- *p* modifies the decay rate and typically falls in the range 0.7–1.5.

According to these equations, the rate of aftershocks decreases quickly with time. The rate of aftershocks is proportional to the inverse of time since the mainshock. Thus whatever the odds of an aftershock are on the first day, the second day will have 1/2 the odds of the first day and the tenth day will have approximately 1/10th the odds of the first day (when p is equal to 1). These patterns describe only the mass behaviour of aftershocks; the actual times, numbers and locations of the aftershocks are 'random', while tending to follow these patterns. As this is an empirical law values of the parameters are obtained by fitting to data after the mainshock occurred and they have no physical basis/meaning.

Bath's Law

The other main law describing aftershocks is known as Bath's Law and this states that the difference in magnitude between a main shock and its largest aftershock is approximately constant, independent of the main shock magnitude, typically 1.1-1.2 on the Moment magnitude scale.

Gutenberg-Richter Law

Aftershock sequences also typically follow the Gutenberg-Richter law of scaling, which refers to the relationship between the magnitude and total number of earthquakes in a region in a given time period. In summary, there are more small aftershocks and fewer large aftershocks.

Impact of Aftershocks

Aftershocks are dangerous because they are usually unpredictable, can be of a large magnitude, and can collapse buildings that are damaged from the mainshock. Bigger earthquakes have more and larger aftershocks and the sequences can last for years or even longer especially when a large event occurs in a seismically quiet area; the New Madrid Seismic Zone, where events still follow Omori's law from the mainshocks of 1811–1812. An aftershock sequence is deemed to have ended when the rate of seismicity drops back to a background level; i.e., no further decay in the number of events with time can be detected.

Foreshocks

Foreshocks have been defined as a tremor of at least 2.5 on the Richter-scale that take place prior to a mainshock tremor of 5.4 or greater on the Richter-scale. Many scientists hope to use foreshocks to predict upcoming earthquakes. In particular, the East Pacific Rise transform faults show foreshock activity before the main seismic event. Reviews of data of past events and their foreshocks showed that they have a low number of aftershocks and high foreshock rates compared to continental strike-slip faults.

EARTHQUAKE SWARM

Earthquake swarms are events where a local area experiences sequences of many earthquakes striking in a relatively short period of time. The length of time used to define the swarm itself varies, but the United States Geological Survey points out that an event may be on the order of days, weeks, or months. They are differentiated from earthquakes succeeded by a series of aftershocks by the observation that no single earthquake in the sequence is obviously the main shock. Earthquake swarms are one of the events typically preceding eruptions of volcanoes.

One example is along the Cerro Prieto fault near Mexicali, BC in Mexico where over 500 quakes and aftershocks hit in February, 2008. Another is a swarm that's been dubbed "The Mogul earthquake sequence" that began in February 2008 near Reno, Nevada and continued for several months, ending in November 2008. Between February and April the swarm produced more than 1,000 quakes of small magnitude, although the largest measured 4.7.

Earthquake Storm

An earthquake storm is a recently proposed theory about earthquakes, where one triggers a series of other large earthquakes—within the same tectonic plate—as the stress transfers along the fault. This is similar to the idea of aftershocks, with the exception that they take place years apart. These series of earthquakes can devastate entire countries or geographical regions. Possible events may have occurred during the end of the Bronze Age, and the latter part of the Roman Empire. It has been suggested that this is what may be occurring in modern day Turkey.

PLATE TECTONICS

The earth beneath our feet is not dead; it is constantly moving, driven by forces deep in its core. Nor is the planet's crust all of one piece; it is composed of numerous plates, which are moving steadily in relation to one another. This movement is responsible for all manner of phenomena, including earthquakes, volcanoes, and the formation of mountains. All these ideas, and many more, are encompassed in the concept of plate tectonics, which is the name for a branch of geologic and geophysical study and for a powerful theory that unites a vast array of ideas. Plate tectonics works hand in hand with several other striking concepts and discoveries, including continental drift and the many changes in Earth's magnetic field that have taken place over its history. No wonder, then, that this idea, developed in the 1960s but based on years of research that preceded that era, is described as "the unifying theory of geology."

TECTONICS AND TECTONISM

The lithosphere is the upper layer of Earth's interior, including the crust and the brittle portion at the top of the mantle. Tectonism is the deformation of the lithosphere, and the term tectonics refers to the study of this deformation, including its causes and effects, most notably mountain building. This deformation is the result of the release and redistribution of energy from Earth's core.

The interior of Earth itself is divided into three major sections: the crust, mantle, and core. The first is the uppermost division of the solid earth, representing less than 1% of its volume and varying in depth from 3 mi. to 37 mi. (5-60 km). Below the crust is the mantle, a thick, dense layer of rock approximately 1,429 mi. (2,300 km) thick. The core itself is even more dense, as illustrated by the fact that it constitutes about 16% of the planet's volume and 32% of its mass. Composed primarily of iron and another, lighter element (possibly sulfur), it is divided between a solid inner core with a radius of about 760 mi. (1,220 km) and a liquid outer core about 1,750 mi. (2,820 km) thick.

Tectonism results from the release and redistribution of energy from Earth's interior. There are two components of this energy: gravity, a function of the enormous mass at the core, and heat from radioactive decay. Differences in mass and heat within the planet's interior, known as pressure gradients, result in the deformation of rocks.

DEFORMATION OF ROCKS

Any attempt to deform an object is referred to as stress, and stress takes many forms, including tension, compression, and shear. Tension acts to stretch a material, whereas compression—a type of stress produced by the action of equal and opposite forces, whose effect is to reduce the length of a material—has the opposite result. (Compression is a form of pressure.) As for shear, this is a kind of stress resulting from equal and opposite forces that do not act along the same line. If a thick, hardbound book is lying flat and one pushes the front cover from the side so that the covers and pages are no longer perfectly aligned, this is an example of shear.

Under the effects of these stresses, rocks may bend, warp, slide, or break. They may even flow, as though they were liquids, or melt and thus truly become liquid. As a result, Earth's interior may manifest faults, or fractures in rocks, as well as folds, or bends in the rock structure. The effects of this activity can be seen on the surface in the form of subsidence, which is a depression in the crust, or uplift, which is the raising of crustal materials. Earthquakes and volcanic eruptions also may result.

There are two basic types of tectonism: orogenesis and epeirogenesis. *Orogenesis* is taken from the Greek words *oros* ("mountain") and *genesis* ("origin") and involves the formation of mountain ranges by means of folding, faulting, and volcanic activity. The Greek word *epeiros* means "mainland," and epeirogenesis takes the form of either uplift or subsidence. Of principal concern in the theory of plate tectonics, as we shall see, is orogenesis, which involves more lateral, as opposed to vertical, movement.

CONTINENTAL DRIFT

If one studies a world map for a period of time, one may notice something interesting about the shape of Africa's west coast and that of South America's east coast: they seem to fit together like pieces of a jigsaw puzzle. Early in the twentieth century, two American geologists, Frank Bursley Taylor and Howard Baker, were among the first scientists to point out this fact. According to Taylor and Baker, Europe, the Americas, and Africa all had been joined at one time. This was an early version of continental drift, a theory concerning the movement of Earth's continents.

Continental drift is based on the idea that the configuration of continents was once different than it is today, that some of the individual landmasses of today once were joined in other continental forms, and that the landmasses later moved to their present locations. Though Taylor and Baker were early proponents, the theory is associated most closely with the German geophysicist and meteorologist Alfred Wegener, who made the case for continental drift in *The Origin of Continents and Oceans*.

PANGAEA, LAURASIA, AND GONDWANALAND

According to Wegener, the continents of today once formed a single supercontinent called Pangaea, from the Greek words *pan* ("all") and *gaea* ("Earth"). Eventually, Pangaea split into two halves, with the northern continent of Laurasia and the southern continent of Gondwanaland, sometimes called Gondwana, separated by the Tethys

Sea. In time, Laurasia split to form North America, the Eurasian landmass with the exception of the Indian subcontinent, and Greenland. Gondwanaland also split, forming the major southern landmasses of the world: Africa, South America, Antarctica, Australia, and India.

The Austrian geologist Eduard Suess and the South African geologist Alexander du Toit, each of whom contributed significantly to continental drift theory, were responsible for the naming of Gondwanaland and Laurasia, respectively. Suess preceded Wegener by many years with his theory of Gondwanaland, named after the Gondwana region of southern India. There he found examples of a fern that, in fossilized form, had been found in all the modern-day constituents of the proposed former continent. Du Toit, Wegener's contemporary, was influenced by continental drift theory and improved on it greatly.

FORMATION OF THE CONTINENTS

Today continental drift theory is accepted widely, in large part owing to the development of plate tectonics, "the unifying theory in geology." We examine the evidence for continental drift, the arguments against it, and the eventual triumph of plate tectonics in the course of this essay. Before going on, however, let us consider briefly the now-accepted timeline of events described by Wegener and others.

About 1,100 million years ago (earth scientists typically abbreviate this by using the notation 1,100 Ma), there was a supercontinent named Rodinia, which predated Pangaea. It split into Laurasia and Gondwanaland, which moved to the northern and southern extremes of the planet, respectively. Starting at about 514 Ma, Laurasia drifted southward until it crashed into Gondwanaland about 425 Ma. Pangaea, surrounded by a vast ocean called Panthalassa, formed approximately 356 Ma.

In the course of Pangaea's formation, what is now North America smashed into northwestern Africa, forming a vast mountain range. Traces of these mountains still can be found on a belt stretching from the southern United States to northern Europe, including the Appalachians. As Pangaea drifted northward and smashed into the ocean floor of Panthalassa, it formed a series of mountain ranges from Alaska to southern South America, including the Rockies and Andes. By about 200 Ma, Pangaea began to break apart, forming a valley that became the Atlantic Ocean. But the separation of the continents was not a "neat" process: today a piece of Gondwanaland lies sunken beneath the eastern United States, far from the other landmasses to which it once was joined.

By about 152 Ma, in the late Jurassic period, the continents as we know them today began to take shape. By about 65 Ma, all the present continents and oceans had been formed for the most part, and India was drifting north, eventually smashing into southern Asia to shape the world's tallest mountains, the Himalayas, the Karakoram Range, and the Hindu Kush. This process is not finished, however, and geologists believe that some 250-300 million years from now, Pangaea will re-form.

EVIDENCE AND ARGUMENTS

As proof of his theory, Wegener cited a wide variety of examples, including the apparent fit between the coastlines of South America and western Africa as well as that of North America and northwestern Africa. He also noted the existence of rocks apparently gouged by glaciers in southern Africa, South America, and India, far from modern-day glacial activity. Fossils in South America matched those in Africa and Australia, as Suess had observed. There were also signs that mountain ranges continued between continents—not only those apparently linking North America and Europe but also ranges that seemed to extend from Argentina to South Africa and Australia.

By measurements conducted over a period of years, Wegener even showed that Greenland was drifting slowly away from Europe, yet his theory met with scorn from the geoscience community of his day. If continents could plow through oceanic rock, some geologists maintained, then they would force up mountains so high that Earth would become imbalanced. As for his claim that matching fossils in widely separated regions confirmed his theory of continental drift, geologists claimed that this could be explained by the existence of land bridges, now sunken, that once had linked those areas. The apparent fit between present-day landmasses could be explained away as coincidence or perhaps as evidence that Earth simply was expanding, with the continents moving away from one another as the planet grew.

PLATE TECTONICS

Though Wegener was right, as it turned out, his theory had one major shortcoming: it provided no explanation of exactly *how* continental drift had occurred. Even if geologists had accepted his claim that the continents are moving, it raised more questions than answers. A continent is a very large thing simply to float away; even an aircraft carrier, which is many millions of times lighter, has to weigh less than the water it displaces, or it would sink like a stone. In any case, Wegener never claimed that continents floated. How, then, did they move?

The answer is plate tectonics, the name both of a theory and of a specialization of tectonics. As an area of study, plate tectonics deals with the large features of the lithosphere and the forces that fashion them. As a theory, it explains the processes that have shaped Earth in terms of plates (large movable segments of the lithosphere) and their movement. Plate tectonics theory brings together aspects of continental drift, seafloor spreading, seismic and volcanic activity, and the structures of Earth's crust to provide a unifying model of Earth's evolution.

THE PIECES COME TOGETHER

In 1962 the United States geologist Harry Hammond Hess introduced a new concept that would prove pivotal to the theory of plate tectonics: seafloor spreading, the idea that seafloors crack open along the crest of mid-ocean ridges and that new seafloor forms in those areas. According to Hess, a new floor forms when molten rock called magma rises up from the asthenosphere, a region of extremely high pressure underlying the lithosphere, where rocks are deformed by enormous stresses. The magma wells up through a crack in a ridge, runs down the sides, and solidifies to form a new floor.

Three years later, the Canadian geologist John Tuzo Wilson coined the term plates to describe the pieces that make up Earth's rigid surface. Separated either by the mid-ocean rifts identified earlier by Heezen or by mountain chains, the plates move with respect to one another. Wilson presented a model for their behaviour and established a global pattern of faults, a sort of map depicting the movable plates. The pieces of a new theory were forming, but as yet it had no name.

That name appeared in 1967, when D. P. Mackenzie of England and R. L. Parker of the United States introduced the term plate tectonics. They maintained that the surface of Earth is divided into six major as well as seven minor movable plates and compared the continents to enormous icebergs—much as Wegener had described them half a century earlier. Subsequent geologic research has indicated that there may be as many as nine major plates and as many as 12 minor ones.

To test these emerging ideas, the U.S. National Science Foundation authorized a research voyage by the vessel *Glomar Challenger* in 1968. On their first cruise, through the Gulf of Mexico and the Atlantic, the *Challenger*'s scientific team collected sediment, fossil, and crust samples that confirmed the basics of seafloor spreading theory. These results led to new questions regarding the reactions between rocks and the heated water surrounding them, spawning new research and necessitating additional voyages. In the years that followed, the

Challenger made more and more cruises, its scientific teams collecting a wealth of evidence for the emerging theory of plate tectonics.

REAL-LIFE APPLICATIONS

Early Evidence of Plate Tectonics

No single person has been as central to plate tectonics as Wegener was to continental drift or as the English naturalist Charles Darwin was to evolution. The roots of plate tectonics lie partly in the observations of Wegener and other proponents of continental drift as well as in several discoveries and observations that began to gather force in the third quarter of the twentieth century.

During World War II, submarine warfare necessitated the development of new navigational technology known as sonar (*SO* und *N* avigation *A* nd *R* anging). Sonar functions much like radar, but instead of using electromagnetic waves, it utilizes ultrasonic, or high-frequency, sound waves projected through water. Sonar made it feasible for geologists to study deep ocean basins after the war, making it possible for the first time in history to map and take samples from large areas beneath the seas. These findings raised many questions, particularly concerning the vast elevation differences beneath the seas.

Ewing and the Mountains Under the Ocean

One of the first earth scientists to notice the curious aspects of underwater geology was the American geologist William Maurice Ewing, who began his work long before the war. He had gained his first experience in a very practical way during the 1920s, as a doctoral student putting himself through school. Working summers with oil exploration teams in the Gulf of Mexico off the coast of Texas had given him a basic understanding of the subject, and in the following decade he went to work exploring the structure of the Atlantic continental shelf and ocean basins.

His work there revealed extremely thick sediments covering what appeared to be high mountainous regions. These findings sharply contradicted earlier ideas about the ocean floor, which depicted it as a flat, featureless plain rather like the sandy-bottomed beaches found in resort areas. Instead, the topography at the bottom of the ocean turned out to be at least as diverse as that of the land above sea level.

Heezen and the Rift Valley

During the 1950s, a team led by another American geologist, Bruce Charles Heezen, worked on developing an overall picture of the ocean basin's topography. Earlier work had identified a mountain range running the length of the Atlantic, but Heezen's team discovered a deep valley down the middle of the chain, running parallel to it. They described it as a rift valley, a long trough bounded by two or more faults, and compared it to a similar valley in eastern Africa.

Around the same time, a group of transatlantic telephone companies asked Heezen to locate areas of possible seismic or earthquake activity in the Atlantic. Phone company officials reasoned that if they could find the areas most likely to experience seismic activity, they could avoid placing their cables in those areas. As it turned out, earthquakes tended to occur in exactly the same region that Heezen and his team had identified as the rift valley.

The Plates and Their Interactions

The most significant plates that make up Earth's surface are as follows:
Selected Major Plates

- North American (almost all of North America and Mexico, along with Greenland and the northwestern quadrant of the Atlantic)
- South American (all of South America and the southwestern quadrant of the Atlantic)
- African (Africa, the southeastern Atlantic, and part of the Indian Ocean)
- Eurasian (Europe and Asia, excluding the Indian subcontinent, along with surrounding ocean areas)
- Indo-Australian (India, much of the Indian Ocean, Australia, and parts of the Indonesian archipelago and New Zealand)
- Antarctic (Antarctica and the Antarctic Ocean)

In addition to these plates, there are several plates that while they are designated as "major" are much smaller: the Philippine, Arabian, Caribbean, Nazca (off the west coast of South America), Cocos (off the west coast of Mexico), and Juan de Fuca (extreme western North America). Japan, one of the most earthquake-prone nations in the world, lies at the nexus of the Philippine, Eurasian, and Pacific plates.

Movement of the Plates

One of the key principles of geology is uniformitarianism: the idea that processes occurring now also occurred in the past. The reverse usually is also true; thus, as we have noted, the plates are still moving, just as they have done for millions of years. Thanks to satellite remote sensing, geologists are able to measure this rate of movement. Not surprisingly, its pace befits the timescale of geologic, as opposed to human, processes: the fastest-moving plates are careening forward at a breathtaking speed of 4 in. (10 cm) per year. The ground beneath Americans' feet is drifting at the rate of 1.2 in. (3 cm) every year, which means that in a hundred years it will have shifted 10 ft. (3 m).

When Plates Interact

Plates interact by moving toward each other (convergence), away from each other (divergence), or past each other (transform motion). Convergence usually is associated with subduction, meaning that one plate is forced down into the mantle and eventually undergoes partial melting. This typically occurs in the ocean, creating a depression known as an oceanic trench. Divergence results in the separation of plates and most often is associated either with seafloor spreading or the formation of rift valleys.

There are three types of plate margins, or boundaries between plates, depending on the two types of crusts that are interacting: oceanic with oceanic, continental with continental, or continental with oceanic. The rift valleys of the Atlantic are an example of an oceanic margin where divergence has occurred, while oceanic convergence is illustrated by a striking example in the Pacific. There, subduction of the Philippine Plate by the Pacific Plate has created the Mariana Trench, which at 36,198 ft. (10,911 m) is the deepest depression on Earth.

When continental plates converge, neither plate subducts; rather, they struggle against each other like two warriors in a fight to the death, buckling, folding, and faulting to create huge mountain ranges. The convergence of the Indo-Australian and Eurasian plates has created the *highest* spots on Earth, in the Himalayas, where Mount Everest (on the Nepal-Indian border) rises to 29,028 ft. (8,848 m). Continental plates also may experience divergence, resulting in the formation of seas. An example is the Red Sea, formed by the divergence of the African and Arabian plates.

Given these facts about the interactions of oceanic and continental plates with each other, what occurs when continental plates meet oceanic ones is no surprise. In this situation, the oceanic plate meeting the continental plate is like a high-school football player squaring off against a National Football League pro tackle. It is no match: the oceanic plate easily subducts. This leads to the formation of a chain of volcanoes along the continental crust, examples being the Cascade Range in the U.S. Pacific Northwest (Juan de Fuca and Pacific plates) or the Andes (South American and Nazca plates).

Transform margins may occur with any combination of oceanic or continental plates and result in the formation of faults and earthquake zones. Where the North American Plate slides against the Pacific Plate along the California coast, it has formed the San Andreas Fault, the source of numerous earthquakes, such as the dramatic San Francisco quakes of 1906 and 1989 and the Los Angeles quake of 1994.

Paleomagnetism

As noted earlier, plate tectonics brings together numerous areas of study in the geologic sciences that developed independently but which came to be seen as having similar roots and explanations. Among these disciplines is paleomagnetism, an area of historical geology devoted to studying the direction and intensity of magnetic fields in the past, as discerned from the residual magnetization of rocks.

Earth has a complex magnetic field whose principal source appears to be the molten iron of the outer core. In fact, the entire planet is like a giant bar magnet, with a north pole and a south pole. It is for this reason that the magnetized material in a compass points north; however, Earth's magnetic north pole is not the same as its geographic north pole. It so happens that magnetic north lies in more or less the same direction as geographic north, but as geologists in the mid-nineteenth century discovered, this has not always been the case.

In 1849 the French physicist Achilles Delesse observed that magnetic minerals tend to line up with the planet's magnetic field, pointing north as though they were compass needles. Nearly 60 years later, however, another French physicist, Bernard Brunhes, noted that in some rocks magnetic materials point *south.* This suggested one of two possibilities: either the planet's magnetic field had reversed itself over time, or the ground containing the magnetized rocks had moved. Both explanations must have seemed far-fetched at the time, but as it turned out, both are correct.

Earth's magnetic field has shifted, meaning that the magnetic north and south poles have changed places many times over the eons. In addition, the magnetic poles have wandered around the southern and northern portions of the globe: for instance, whereas magnetic north today lies in the frozen islands to the north of Canada, at about 300 Ma it was located in eastern Siberia. The movement of magnetic rocks on Earth's surface, however, has turned out to be too great to be explained either by magnetic shifts or by regional wandering of the poles. This is where plate tectonics and paleomagnetism come together.

Confirmation of Plate Tectonic Theory

Rocks in Alaska have magnetic materials aligned in such a way that they once must have been at or near the equator. In addition, the orientation of magnetic materials on South America's east coast shows an affinity with that of similar materials on the west coast of Africa. In both cases, continental drift, with its driving mechanism of plate tectonics, seems the only reasonable explanation.

Thus, paleomagnetic studies have served to confirm the ideas of continental drift and plate tectonics, while research conducted at sea bolsters seafloor spreading theory. Using devices called magnetometers, geologists have found that the orientation of magnetic minerals on one side of a rift mirrors that of materials on the other side. This suggests that the new rock on either side of the rift was formed simultaneously, as seafloor spreading theory indicates.

Earthquakes and Volcanoes

Several findings relating to earthquakes and volcanic activity also can be explained by plate tectonics. If one follows news stories of earthquakes, one may begin to wonder why such places as California or Japan have so many quakes, whereas the northeastern United States or western Europe have so few. The fact is that earthquakes occur along belts, and the vast majority of these belts coincide with the boundaries between Earth's major tectonic plates.

The same is true of volcanoes, and it is no mistake that places famous for earthquakes—the Philippines, say, or Italy—often also are known for their volcanoes. Although they are located near the centre of the Pacific Plate, the islands of Hawaii are subject to plate movement, which has helped generate the volcanoes that gave those islands their origin. At the southern end of the island chain, many volcanoes are still active, while those at the northern end tend to be dormant. The reason is that the Pacific Plate as a whole is moving northward over a stationary lava source in the mantle below Hawaii. The southern islands remain poised above that source, while the northern islands have moved away from it.

The Oceanic and Continental Crusts

Given what we have seen about continental drift and seafloor spreading, it should come as no surprise to learn that, generally speaking, the deeper one goes in the ocean, the newer the crust. Specifically, the crust is youngest near the centre of ocean basins and particularly along mid-ocean ridges, or submarine mountain ridges where new seafloor is created by seafloor spreading.

It also should not be surprising to learn that oceanic and continental crusts differ both in thickness and in composition. Basalt, an igneous rock (rock formed from the cooling of magma), makes up the preponderance of ocean crust, whereas much of the continental crust is made up of granite, another variety of igneous rock. Whereas the ocean crust is thin, generally 3-6 mi. (5-10 km) in depth, the continental crust ranges in thickness from 12.5-55 mi. (20-90 km). This results in a difference in thickness for the lithosphere, which is only about 60 mi. (100 km) thick beneath the oceans but about 2.5 times as thick—150 mi. (250 km)—under the continents.

Chapter 10

Hydrogeology

Hydrogeology is the area of geology that deals with the distribution and movement of groundwater in the soil and rocks of the Earth's crust, (commonly in aquifers). The term geohydrology is often used interchangeably. Some make the minor distinction between a hydrologist or engineer applying themselves to geology (geohydrology), and a geologist applying themselves to hydrology (hydrogeology).

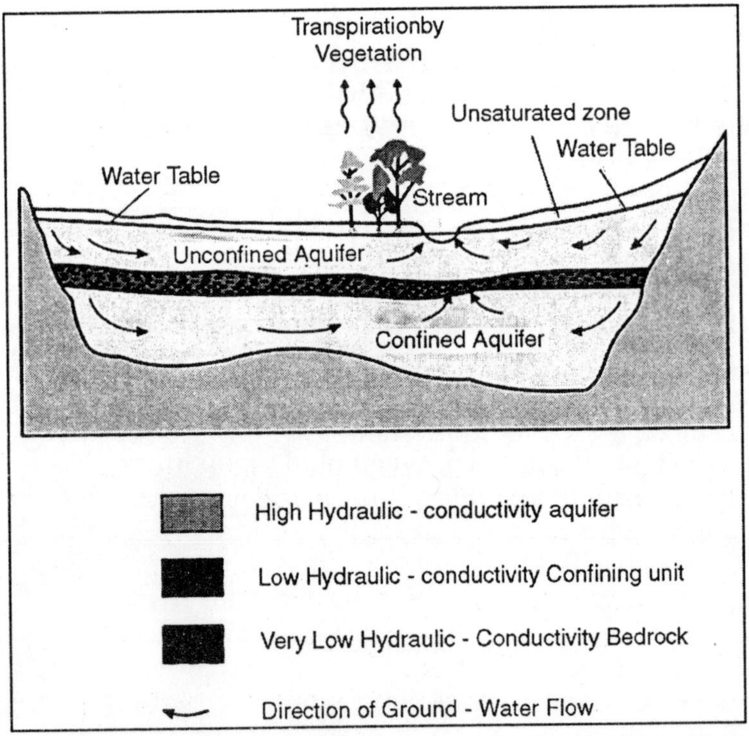

Hydrogeology (like most earth sciences) is an interdisciplinary subject; it can be difficult to account fully for the chemical, physical, biological and even legal interactions between soil, water, nature and society. The study of the interaction between groundwater movement and geology can be quite complex. Groundwater does not always flow in the subsurface down-hill following the surface topography; groundwater follows pressure gradients (flow from high pressure gradient to low) often following fractures and conduits in circuitous paths. Taking into account the interplay of the different facets of a multi-component system often requires knowledge in several diverse fields at both the experimental and theoretical levels. This being said, the following is a more traditional (reductionist viewpoint) introduction to the methods and nomenclature of saturated subsurface hydrology, or simply hydrogeology.

HYDROGEOLOGY RELATED WITH OTHER FIELDS

Hydrogeology, as stated above, is a branch of the earth sciences dealing with the flow of water through aquifers and other shallow porous media (typically less than 450 m or 1,500 ft below the land surface.) The very shallow flow of water in the subsurface (the upper 3 m or 10 ft) is pertinent to the fields of soil science, agriculture and civil engineering, as well as to hydrogeology. The general flow of fluids (water, hydrocarbons, geothermal fluids, etc.) in deeper formations is also a concern of geologists, geophysicists and petroleum geologists. Groundwater is a slow-moving, viscous fluid (with a Reynolds number less than unity); many of the empirically derived laws of groundwater flow can be alternately derived in fluid mechanics from the special case of Stokes flow (viscosity and pressure terms, but no inertial term).

The mathematical relationships used to describe the flow of water through porous media are the diffusion and Laplace equations, which have applications in many diverse fields. Steady groundwater flow (Laplace equation) has been simulated using electrical, elastic and heat conduction analogies. Transient groundwater flow is analogous to the diffusion of heat in a solid, therefore some solutions to hydrological problems have been adapted from heat transfer literature.

Traditionally, the movement of groundwater has been studied separately from surface water, climatology, and even the chemical and microbiological aspects of hydrogeology (the processes are uncoupled). As the field of hydrogeology matures, the strong interactions between groundwater, surface water, water chemistry, soil moisture and even climate are becoming more clear.

HYDRAULIC HEAD

Hydraulic head or piezometric head is a specific measurement of water pressure above a geodetic datum. It is usually measured as a water surface elevation, expressed in units of length, at the entrance (or bottom) of a piezometer. In an aquifer, it can be calculated from the depth to water in a piezometric well (a specialized water well), and given information of the piezometer's elevation and screen depth. Hydraulic head can similarly be measured in a column of water using a standpipe piezometer by measuring the height of the water surface in the tube relative to a common datum. The hydraulic head can be used to determine a *hydraulic gradient* between two or more points.

"HEAD" IN FLUID DYNAMICS

In fluid dynamics, head is a concept that relates the energy in an incompressible fluid to the height of an equivalent static column of that fluid.

From Bernoulli's Principle, the total energy at a given point in a fluid is the energy associated with the movement of the fluid, plus energy from pressure in the fluid, plus energy from the height of the fluid relative to an arbitrary datum. Head is expressed in units of height such as meters and feet.

The static head of a pump is the maximum height (pressure) it can deliver. The capability of the pump can be read from its Q-H curve (flow vs. height).

Head is equal to the fluid's energy per unit weight. Head is useful in specifying centrifugal pumps because their pumping characteristics tend to be independent of the fluid's density.

There are four types of head used to calculate the total head in and out of a pump:

- *Velocity head:* Is due to the bulk motion of a fluid (kinetic energy).
- *Elevation head:* Is due to the fluid's weight, the gravitational force acting on a column of fluid.
- *Pressure head:* Is due to the static pressure, the internal molecular motion of a fluid that exerts a force on its container.
- *Resistance head* (or *friction head* or Head Loss): Is due to the frictional forces acting against a fluid's motion by the container.

COMPONENTS OF HYDRAULIC HEAD

A mass free falling from an elevation (in a vacuum) will reach a speed

$$v = \sqrt{2gz},$$

when arriving at elevation $z = 0$, or when we rearrange it as a head:

$$h = \frac{v^2}{2g}$$

where, g is the acceleration due to gravity.

The term $\frac{v^2}{2g}$ is called the *velocity head*, expressed as a length measurement. In a flowing fluid, it represents the energy of the fluid due to its bulk motion.

The velocity head, or total hydraulic head of a fluid is composed of *pressure head* and *elevation head*.

The pressure head is the equivalent gauge pressure of a column of water at the base of the piezometer, and the elevation head is the relative potential energy in terms of an elevation. The Head Equation, a simplified form of the Bernoulli Principle for incompressible fluids, can be expressed as:

$$h = \psi + z$$

where

h is the hydraulic head (Length in m or ft), also known as the velocity head. ψ is the pressure head, in terms of the elevation difference of the water column relative to the piezometer bottom (Length in m or ft), and z is the elevation at the piezometer bottom (Length in m or ft).

In an example with a 400 m deep piezometer, with an elevation of 1000 m, and a depth to water of 100 m: $z = 600$ m, $\psi = 300$ m, and $h = 900$ m.

The pressure head can be expressed as:

$$\psi = \frac{P}{\gamma} = \frac{P}{pq}$$

where

P is the gauge pressure (Force per unit area, often Pa or psi), γ is the unit weight of water (Force per unit volume, typically $N \cdot m^{-3}$ or lbf/ft^3), ρ is the density of the water (Mass per unit volume, frequently $kg \cdot m^{-3}$), and g is the gravitational acceleration (velocity change per unit time, often $m \cdot s^{-2}$)

FRESH WATER HEAD

The pressure head is dependent on the density of water, which can vary depending on both the temperature and chemical composition (salinity, in particular). This means that the hydraulic head calculation is dependent on the density of the water within the piezometer. If one or more hydraulic head measurements are to be compared, they need to be standardized, usually to their *fresh water head,* which can be calculated as:

$$h_{fw} = \psi = \frac{P}{\rho fw} + z$$

where

h_{fw} is the fresh water head (Length, measured in m or ft), and

ρfw is the density of fresh water (Mass per unit volume, typically in kg·m^{-3})

HYDRAULIC GRADIENT

The hydraulic gradient is a vector gradient between two or more hydraulic head measurements over the length of the flow path. It is also called the Darcy slope, since it determines the quantity of a Darcy flux, or discharge. A dimensionless hydraulic gradient can be calculated between two piezometers as:

$$i = \frac{dh}{dl} = \frac{h_2 - h_1}{\text{length}}$$

where

i is the hydraulic gradient (dimensionless),

dh is the difference between two hydraulic heads (Length, usually in m or ft), and

dl is the flow path length between the two piezometers (Length, usually in m or ft)

The hydraulic gradient can be expressed in vector notation, using the del operator. This requires a hydraulic head field, which can only be practically obtained from a numerical model, such as MODFLOW. In Cartesian coordinates, this can be expressed as:

$$\nabla h = \left(\frac{\partial h}{\partial x}, \frac{\partial h}{\partial y}, \frac{\partial h}{\partial z} \right) = \frac{\partial h}{\partial x} i + \frac{\partial h}{\partial y} j + \frac{\partial h}{\partial z} k$$

This vector describes the direction of the groundwater flow, where negative values indicate flow along the dimension, and zero indicates *no flow*. As with any other example in physics, energy must flow from high to low, which is why the flow is in the negative gradient. This vector can be used in conjunction with Darcy's law and a tensor of hydraulic conductivity to determine the flux of water in three dimensions.

Hydraulic head in Groundwater

The distribution of hydraulic head through an aquifer determines where groundwater will flow. In a hydrostatic example, where the hydraulic head is constant, there is no flow. However, if there is a difference in hydraulic head from the top to bottom due to draining from the bottom, the water will flow downward, due to the difference in head, also called the hydraulic gradient.

ATMOSPHERIC PRESSURE

Even though it is convention to use gauge pressure in the calculation of hydraulic head, it is more correct to use total pressure (gauge pressure + atmospheric pressure), since this is truly what drives groundwater flow. Often detailed observations of barometric pressure are not available at each well through time, so this is often disregarded (contributing to large errors at locations where hydraulic gradients are low or the angle between wells is acute.)

Relation between heads for a hydrostatic case and a downward flow case.

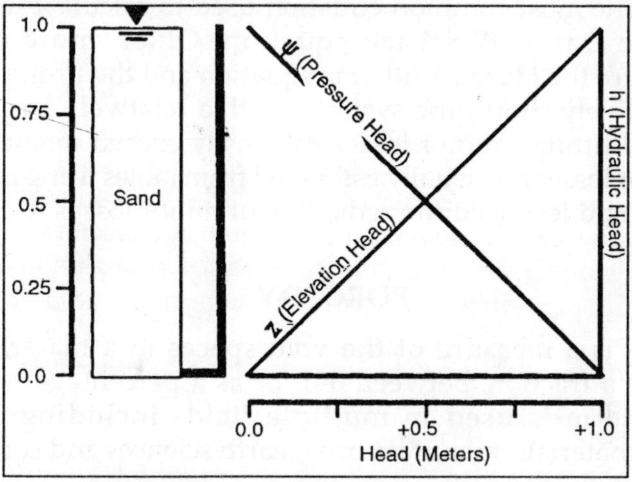

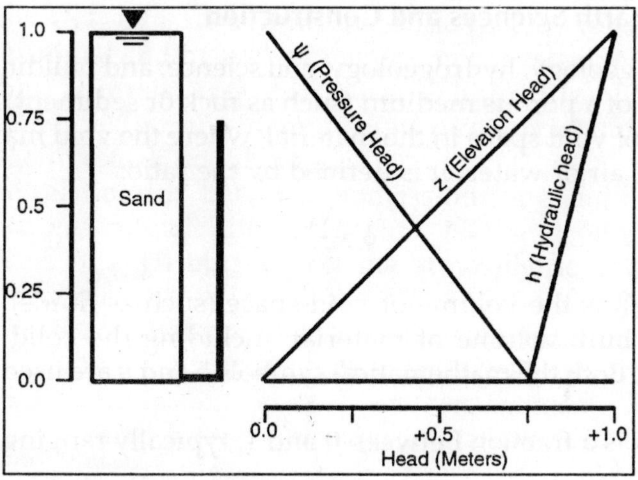

The effects of changes in atmospheric pressure upon water levels observed in wells has been known for many years. The effect is a direct one, an increase in atmospheric pressure is an increase in load on the water in the aquifer, which increases the depth to water (lowers the water level elevation). Pascal first qualitatively observed these effects in the 1600s, and they were more rigorously described by the soil physicist Edgar Buckingham (working for the USDA) using air flow models in 1907.

HEAD LOSS

In any real moving fluid, energy is dissipated due to friction, as well as turbulence, unless the flow is laminar. Head loss is divided into two main categories, "major losses" associated with energy loss per length of pipe, and "minor losses" associated with bends, fittings, valves, etc. The most common equation used to calculate major head losses is the Darcy–Weisbach equation. Older, more empirical approaches are the Hazen-Williams equation and the Prony equation.

For relatively short pipe systems, with a relatively large number of bends and fittings, minor losses can easily exceed major losses. In design, minor losses are usually estimated from tables using coefficients or a simpler and less accurate reduction of minor losses to equivalent length of pipe.

POROSITY

Porosity is a measure of the void spaces in a material, and is measured as a fraction, between 0–1, or as a percentage between 0–100%. The term is used in multiple fields including ceramics, metallurgy, materials, manufacturing, earth sciences and construction.

Porosity in Earth Sciences and Construction

Used in geology, hydrogeology, soil science, and building science, the porosity of a porous medium (such as rock or sediment) describes the fraction of void space in the material, where the void may contain, for example, air or water. It is defined by the ratio:

$$\phi = \frac{V_V}{V_T}$$

where V_V is the volume of void-space (such as fluids) and V_T is the total or bulk volume of material, including the solid and void components. Both the mathematical symbols ö and n are used to denote porosity.

Porosity is a fraction between 0 and 1, typically ranging from less than 0.01 for solid granite to more than 0.5 for peat and clay, although it may also be represented in per cent terms by multiplying the fraction by 100.

The porosity of a rock, or sedimentary layer, is an important consideration when attempting to evaluate the potential volume of water or hydrocarbons it may contain. Sedimentary porosities are a complex function of many factors, including but not limited to: rate of burial, depth of burial, the nature of the connate fluids, the nature of overlying sediments (which may impede fluid expulsion). One commonly used relationship between porosity and depth is given by the Athy (1930) equation:

$$\phi(z)=\phi_0 e^{-kz}$$

where φ_0 is the surface porosity, k is the compaction coefficient (m^{-1}) and z is depth (m).

A value for porosity can alternatively be calculated from the bulk density ρ_{bulk} and particle density $\rho_{particle}$:

$$\phi=1-\frac{\rho_{bulk}}{\rho_{particle}}$$

Normal particle density is assumed to be approximately 2.65 g/cm³, although a better estimation can be obtained by examining the lithology of the particles.

Porosity and Hydraulic Conductivity

Porosity is indirectly related to hydraulic conductivity; for two similar sandy aquifers, the one with a higher porosity will typically have a higher hydraulic conductivity (more open area for the flow of water), but there are many complications to this relationship. Clays, which typically have very low hydraulic conductivity also have very high porosities (due to the structured nature of clay minerals), which means clays can hold a large volume of water per volume of bulk material, but they do not release water rapidly.

Sorting and Porosity

Well sorted (grains of approximately all one size) materials have higher porosity than similarly sized poorly sorted materials (where smaller particles fill the gaps between larger particles). The graphic illustrates how some smaller grains can effectively fill the pores (where all water flow takes place), drastically reducing porosity and hydraulic conductivity, while only being a small fraction of the total volume of the material.

Fig. Effects of Sorting on Alluvial Porosity

Porosity of Rocks

Consolidated rocks (e.g. sandstone, shale, granite or limestone) potentially have more complex "dual" porosities, as compared with alluvial sediment. The rock itself may have a certain (low) porosity, and the fractures (cracks and joints), or dissolution features may create a second (higher) porosity. The interaction of these porosities is complex and often makes simple models highly inaccurate.

Porosity of Soil

Porosity of surface soil typically decreases as particle size increases. This is due to soil aggregate formation in finer textured surface soils when subject to soil biological processes. Aggregation involves particulate adhesion and higher resistance to compaction. Typical bulk density of sandy soil is between 1.5 and 1.7 g/cm³. This calculates to a porosity between 0.43 and 0.36. Typical bulk density of clay soil is between 1.1 and 1.3 g/cm³. This calculates to a porosity between 0.58 and 0.51. This seems counterintuitive because clay soils are termed *heavy*, implying *lower* porosity. Heavy apparently refers to a gravitational moisture content effect in combination with terminology that harkens back to the relative force required to pull a tillage implement through the clayey soil at field moisture content as compared to sand.

Porosity of subsurface soil is lower than in surface soil due to compaction by gravity. Porosity of 0.20 is considered normal for unsorted gravel size material at depths below the biomantle. Porosity in finer material below the aggregating influence of pedogenesis can be expected to approximate this value.

Soil porosity is complex. Traditional models regard porosity as continuous. This fails to account for anomalous features and produces only approximate results. Furthermore it cannot help model the influence of environmental factors which affect pore geometry. A number of more complex models have been proposed, including fractals, bubble theory, cracking theory, Boolean grain process, packed sphere, and numerous other models.

Types of Geologic Porosities

- *Primary porosity:* The main or original porosity system in a rock or unconfined alluvial deposit.
- *Secondary porosity:* A subsequent or separate porosity system in a rock, often enhancing overall porosity of a rock. This can be a result of chemical leeching of minerals or the generation of a fracture system. This can replace the primary porosity or coexist with it.
- *Fracture porosity:* This is porosity associated with a fracture system or faulting. This can create secondary porosity in rocks that otherwise would not be reservoirs for hydrocarbons due to their primary porosity being destroyed (for example due to depth of burial) or of a rock type not normally considered a reservoir (for example igneous intrusions or metasediments).
- *Vuggy porosity:* This is secondary porosity generated by dissolution of large features (such as macrofossils) in carbonate rocks leaving large holes, vugs, or even caves.
- *Effective porosity (also called open porosity):* Refers to the fraction of the total volume in which fluid flow is effectively taking place (this excludes dead-end pores or non-connected cavities). This is very important for groundwater and petroleum flow, as well as for solute transport.
- *Dual porosity:* Refers to the conceptual idea that there are two overlapping reservoirs which interact. In fractured rock aquifers, the rock mass and fractures are often simulated as being two overlapping but distinct bodies. Delayed yield, and leaky aquifer flow solutions are both mathematically similar solutions to that obtained for dual porosity; in all three cases water comes from two mathematically different reservoirs (whether or not they are physically different).
- *Macro porosity:* Refers to pores greater than 50 nm in diameter. Flow through macropores is described by bulk diffusion.
- *Meso porosity:* Refers to pores greater than 2 nm and less than 50 nm in diameter. Flow through mesopores is described by Knudsen diffusion.
- *Micro porosity:* Refers to pores smaller than 2 nm in diameter. Movement in micropores is by activated diffusion.

Measuring Porosity

Several methods can be employed to measure porosity, including the volume/density method (pore volume = total volume - material volume), water saturation method (pore volume = total volume of water - unsaturated water), water evaporation method (pore volume in cubic centimeters = weight of saturated sample in grams - weight of dried sample in grams), mercury intrusion porosimetry (several non-mercury intrusion techniques have been developed due to toxicological concerns, and the fact that mercury tends to form amalgams with several metals/alloys), and nitrogen gas adsorption (nitrogen gas adsorption in pores is measured either by volume or weight. This technique is suitable for materials with very fine pores).

HYDRAULIC CONDUCTIVITY

Hydraulic conductivity, symbolically represented as K, is a property of vascular plants, soil or rock, that describes the ease with which water can move through pore spaces or fractures. It depends on the intrinsic permeability of the material and on the degree of saturation. Saturated hydraulic conductivity, Ksat, describes water movement through saturated media. One application of it is the Starling equation, which calculates flow across walls of capillaries.

DERIVATION THROUGH DARCY'S LAW

Hydraulic conductivity is the proportionality constant in Darcy's law, which relates the amount of water which will flow through a unit cross-sectional area of aquifer under a unit gradient of hydraulic head. It is analogous to the thermal conductivity of materials in heat conduction, or the inverse of resistivity in electrical circuits. The hydraulic conductivity (K — the English letter "kay") is specific to the flow of a certain fluid (typically water, sometimes oil or air); intrinsic permeability (ê — the Greek letter "kappa") is a parameter of a porous media which is independent of the fluid. This means that, for example, K will increase if the water in a porous medium is heated (reducing the viscosity of the water), but ê will remain constant. The two are related through the following equation:

$$K = \frac{\kappa\gamma}{\mu}$$

where
K is the hydraulic conductivity [LT^{-1} or $m\ s^{-1}$]; κ
is the intrinsic permeability of the material [L^2 or m^2]; γ
is the specific weight of water [$ML^{-2}T^{-2}$ or $N\ m^{-3}$], and; μ
is the dynamic viscosity of water [$ML^{-1}T^{-1}$ or $kg\ m^{-1}\ s^{-1}$].

ESTIMATION OF HYDRAULIC CONDUCTIVITY

Direct Estimation

Hydraulic conductivity can be measured by applying Darcy's law on the material. Such experiments can be conducted by creating a hydraulic gradient between two points, and measuring the flow rate.

Empirical Estimation

Shepherd derived an empirical formula for approximating hydraulic conductivity from grain size analyses:

$$K = a(D_{10})^b$$

where

a and b are empirically derived terms based on the soil type, and

D_{10} is the diameter of the 10 percentile grain size of the material

Note: Shepherd's Figure clearly shows the use of d_{50}, not d_{10}, measured in mm. Therefore the equation should be $K = a(d_{10})^b$. His figure shows different lines for materials of different types, based on analysis of data from others with d_{50} up to 10 mm.

Pedotransfer Function

A pedotransfer function (PTF) is a specialized empirical estimation method, used primarily in the soil sciences, however has increasing use in hydrogeology· There are many different PTF methods, however, they all attempt to determine soil properties, such as hydraulic conductivity, given several measured soil properties, such as soil particle size, and bulk density.

Experimental Approach

There are relatively simple and inexpensive laboratory tests that may be run to determine the hydraulic conductivity of a soil: constant-head method and falling-head method.

Constant-Head Method

The constant-head method is typically used on granular soil. This procedure allows water to move through the soil under a steady state head condition while the quantity (volume) of water flowing through the soil specimen is measured over a period of time. By knowing the quantity Q of water measured, length L of specimen, cross-sectional area A of the specimen, time t required for the quantity of water Q to be discharged, and head h, the hydraulic conductivity can be calculated:

$$Q = Avt$$

$$\text{Using Darcy's Law, } v = Ki,$$

$$\text{yields } Q = \frac{AKht}{L}$$

Solving for *K* gives,

$$K = \frac{QL}{Ath}$$

Falling-Head Method

The falling-head method is very similar to the constant head methods in its initial setup; however, the advantage to the falling-head method is that can be used for both fine-grained and coarse-grained soils. The soil sample is first saturated under a specific head condition. The water is then allowed to flow through the soil without maintaining a constant pressure head.

$$K = \frac{2.3aL}{At} \log\left(\frac{h_1}{h_2}\right)$$

Transmissivity

The transmissivity, T, *of an aquifer is a measure of how much water can be transmitted horizontally, such as to a pumping well:*

$$T = K_s b$$

Transmissivity is directly proportional to average permeability (K_s) and aquifer thickness (*b*). For a confined aquifer, this remains constant, as the saturated thickness remains constant. The aquifer thickness of an unconfined aquifer is from the base of the aquifer (or the top of the aquitard) to the water table. The water table can fluctuate, which changes the transmissivity of the unconfined aquifer. This may provide positive feedback of a pumping well that is pumping more than can be provided by the aquifer, where the transmissivity drops as the well pumps, thus eventually reducing the aquifer to the height of the pumping well screen.

Transmissivity should not be confused with similar word transmittance (used in optics), which means fraction of incident light that passes through a sample.

Relative Properties

Because of their high porosity and permeability, sand and gravel aquifers have higher hydraulic conductivity than clay or unfractured granite aquifers. Sand or gravel aquifers would thus be easier to extract water from (e.g., using a pumping well) because of their high transmissivity, compared to clay or unfractured bedrock aquifers.

Hydraulic conductivity has units with dimensions of length per time (e.g., m/s, ft/day and (gal/day)/ft^2); transmissivity then has units with dimensions of length squared per time. The following table gives some typical ranges (illustrating the many orders of magnitude which are likely) for K values.

Hydraulic conductivity (K) is one of the most complex and important of the properties of aquifers in hydrogeology as the values found in nature:

- Range over many orders of magnitude (the distribution is often considered to be lognormal),
- Vary a large amount through space (sometimes considered to be randomly spatially distributed, or stochastic in nature),
- Are directional (in general K is a symmetric second-rank tensor; e.g., vertical K values can be several orders of magnitude smaller than horizontal K values),
- Are scale dependent (testing a m^3 of aquifer will generally produce different results than a similar test on only a cm^3 sample of the same aquifer),
- Must be determined indirectly through field pumping tests, laboratory column flow tests or inverse computer simulation, (sometimes also from grain size analyses), and
- Are very dependent (in a non-linear way) on the water content, which makes solving the unsaturated flow equation difficult. In fact, the variably saturated K for a single material varies over a wider range than the saturated K values for all types of materials.

SPECIFIC STORAGE

Specific storage (S_s), storativity (S), specific yield (S_y) and specific capacity are material physical properties that characterize the capacity of an aquifer to release groundwater from storage in response to a decline in hydraulic head. For that reason they are sometimes referred to as "storage properties". In the field of hydrogeology, these properties are often determined using some combination of field hydraulic tests (e.g., aquifer tests) and laboratory tests on aquifer material samples.

The specific storage: Is the amount of water that a portion of an aquifer releases from storage, per unit mass or volume of aquifer, per unit change in hydraulic head, while remaining fully saturated.

Mass specific storage: Is the mass of water than an aquifer releases from storage, per mass of aquifer, per unit decline in hydraulic head:

$$(S_S)_m = \frac{1}{m_a}\frac{\Delta m_m}{\Delta h}$$

where

$(S_s)_m$ is the mass specific storage ($[L^{-1}]$);

m_a is the mass of that portion of the aquifer from which the water is released ([M]);

Δm_w is the mass of water released from storage ([M]); and

Δh is the decline in hydraulic head ([L]).

Volumetric specific storage (or volume specific storage) is the volume of water that an aquifer releases from storage, per volume of aquifer, per unit decline in hydraulic head:

$$(S_S) = \frac{1}{V_a}\frac{\Delta Vw}{\Delta h}$$

where

S_s is the volumetric specific storage ($[L^{-1}]$);

V_a is the bulk volume of that portion of the aquifer from which the water is released ($[L^3]$);

ΔV_w is the volume of water released from storage ($[L^3]$); and

Δh is the decline in hydraulic head ([L]).

In hydrogeology, volumetric specific storage is much more commonly encountered than mass specific storage. Consequently, the term specific storage generally refers to volumetric specific storage.

In terms of measurable physical properties, specific storage can be expressed as

$$(S_S) = \gamma(\beta_P + n \cdot \beta_w)$$

where γ

is the specific weight of water ($N \bullet m^{-3}$ or $[ML^{-2}T^{-2}]$)

n is the porosity of the material (dimensionless ratio between 0 and 1)

β_p is the compressibility of the bulk aquifer material (m^2N^{-1} or $[LM^{-1}T^2]$), and

β_w is the compressibility of water (m^2N^{-1} or $[LM^{-1}T^2]$)

The compressibility terms relate a given change in stress to a change in volume (a strain). These two terms can be defined as:

$$\beta_P = -\frac{dV_t}{d\sigma_e}\frac{1}{Vt}$$

$$\beta_w = -\frac{dV_w}{d_P}\frac{1}{V_w}$$

where

σ_e is the effective stress (N or $[MLT^{-2}]$)

These equations relate a change in total or water volume (V_t or V_w) per change in applied stress (effective stress — σ_e or pore pressure — p) per unit volume. The compressibilities (and therefore also S_s) can be estimated from laboratory consolidation tests (in an apparatus called a consolidometer), using the consolidation theory of soil mechanics.

Storativity

Storativity is the volume of water released from storage per unit decline in hydraulic head in the aquifer, per unit area of the aquifer, or:

$$S = -\frac{dV_w}{dh}\frac{1}{A}$$

Storativity is the vertically integrated specific storage value for a confined aquifer or aquitard. For a confined homogeneous aquifer or aquitard they are simply related by:

$$S = S_s b$$

where b is the thickness of aquifer. Storativity is a dimensionless quantity, and ranges between 0 and the effective porosity of the aquifer; although for confined aquifers, this number is usually much less than 0.01.

The storage coefficient of an unconfined aquifer is approximately equal to the specific yield, S_y, since the release from specific storage, S_s is typically orders of magnitude less.

Specific Yield

Specific yield, also known as the drainable porosity, is a ratio, less than or equal to the effective porosity, indicating the volumetric fraction of the bulk aquifer volume that a given aquifer will yield when all the water is allowed to drain out of it under the forces of gravity:

where

$$S_y - \frac{d_{wd}}{V_T}$$

V_{wd} is the volume of water drained, and

V_T is the total rock or material volume

It is primarily used for unconfined aquifers, since the elastic storage component, S_s, is relatively small and usually has an insignificant contribution. Specific yield can be close to effective porosity, but there are several subtle things which make this value more complicated than it seems. Some water always remains in the formation, even after drainage; it clings to the grains of sand and clay in the formation. Also, the value of specific yield may not be fully realized until very large times, due to complications caused by unsaturated flow.

DARCY'S LAW

In fluid dynamics and hydrology, Darcy's law is a phenomenologically derived constitutive equation that describes the flow of a fluid through a porous medium. The law was formulated by Henry Darcy based on the results of experiments on the flow of water through beds of sand. It also forms the scientific basis of fluid permeability used in the earth sciences.

Although Darcy's law (an expression of conservation of momentum) was determined experimentally by Henri Darcy, it has since been derived from the Navier-Stokes equations via homogenization. It is analogous to Fourier's law in the field of heat conduction, Ohm's law in the field of electrical networks, or Fick's law in diffusion theory.

One application of Darcy's law is to water flow through an aquifer. Darcy's law along with the equation of conservation of mass are equivalent to the groundwater flow equation, one of the basic relationships of hydrogeology. Darcy's law is also used to describe oil, water, and gas flows through petroleum reservoirs.

Description

Darcy's law is a simple proportional relationship between the instantaneous discharge rate through a porous medium, the viscosity of the fluid and the pressure drop over a given distance.

$$Q = \frac{-\kappa A}{\mu} \frac{(P_b - P_a)}{L}$$

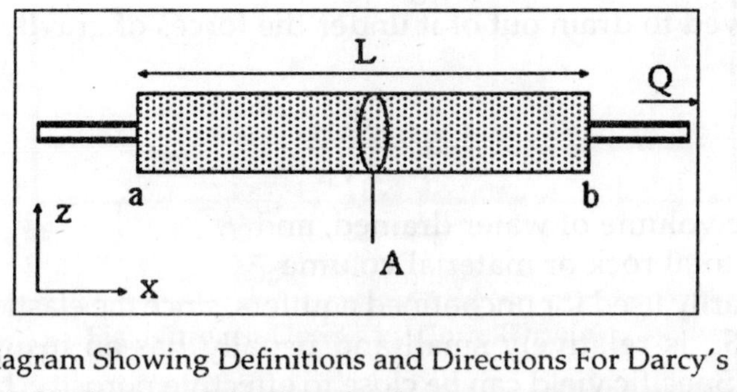

Fig. Diagram Showing Definitions and Directions For Darcy's Law.

The total discharge, *Q* (units of volume per time, e.g., m^3/s) is equal to the product of the permeability (ê units of area, e.g. m^2) of the medium, the cross-sectional area (*A*) to flow, and the pressure drop

(P_b " P_a), all divided by the dynamic viscosity ì (in SI units e.g. kg/(m·s) or Pa·s), and the length *L* the pressure drop is taking place over. The negative sign is needed because fluids flow from high pressure to low pressure. So if the change in pressure is negative (in the *x*-direction) then the flow will be positive (in the *x*-direction). Dividing both sides of the equation by the area and using more general notation leads to

$$q = \frac{-\kappa}{\mu} \nabla P$$

where *q* is the flux (discharge per unit area, with units of length per time, m/s) and ∇P is the pressure gradient vector. This value of flux, often referred to as the Darcy flux, is not the velocity which the water traveling through the pores is experiencing.

The pore velocity (*v*) is related to the Darcy flux (*q*) by the porosity (φ). The flux is divided by porosity to account for the fact that only a fraction of the total formation volume is available for flow. The pore velocity would be the velocity a conservative tracer would experience if carried by the fluid through the formation.

$$v = \frac{q}{\phi}$$

In 3D

In three dimensions, gravity must be accounted for, as the flow is not affected by the vertical pressure drop caused by gravity when assuming hydrostatic conditions. The solution is to subtract the gravitational pressure drop from the existing pressure drop in order to express the resulting flow,

$$q = \frac{-\kappa}{\mu}(\nabla P - \rho g \hat{e}_z)$$

where the flux q is now a vector quantity, κ is a tensor of permeability, ∇ is the gradient operator in 3D, *g* is the acceleration due to gravity, $\hat{e}_z$ is the unit vector in the vertical direction, pointing downwards and ñ is the density.

Effects of anisotropy in three dimensions are addressed using a symmetric second-order tensor of permeability:

$$q = \frac{-\kappa}{\mu}(\nabla P - \rho g \hat{e}_z)\kappa = \begin{bmatrix} \kappa_{xx} & \kappa_{xy} & \kappa_{xz} \\ \kappa_{yx} & \kappa_{yy} & \kappa_{yz} \\ \kappa_{zx} & \kappa_{zy} & \kappa_{zz} \end{bmatrix}$$

where the magnitudes of permeability in the *x*, *y*, and *z* component directions are specified. Since this a symmetric matrix, there are *at most* six unique values. If the permeability is isotropic (equal magnitude in all directions), then the diagonal values are equal, $\kappa_{xx} = \kappa_{yy} = \kappa_{zz} > 0,$, while all other components are 0. The permeability tensor can be interpreted through an evaluation of the relative magnitudes of each component. For example, rock with highly permeable vertical fractures aligned in the *x*-direction will have relatively higher values for than other component values.

Assumptions

Darcy's law is a simple mathematical statement which neatly summarizes several familiar properties that groundwater flowing in aquifers exhibits, including:

- If there is no pressure gradient over a distance, no flow occurs (this is hydrostatic conditions),
- If there is a pressure gradient, flow will occur from high pressure towards low pressure (opposite the direction of increasing gradient - hence the negative sign in Darcy's law),
- The greater the pressure gradient (through the same formation material), the greater the discharge rate, and
- The discharge rate of fluid will often be different — through different formation materials (or even through the same material, in a different direction) — even if the same pressure gradient exists in both cases.

A graphical illustration of the use of the steady-state groundwater flow equation (based on Darcy's law and the conservation of mass) is in the construction of flownets, to quantify the amount of groundwater flowing under a dam.

Darcy's law is only valid for slow, viscous flow; fortunately, most groundwater flow cases fall in this category. Typically any flow with a Reynolds number less than one is clearly laminar, and it would be valid to apply Darcy's law. Experimental tests have shown that flow regimes with values of Reynolds number up to 10 may still be Darcian. Reynolds number (a dimensionless parameter) for porous media flow is typically expressed as

$$\mathrm{Re} = \frac{\rho v d_{30}}{\mu}$$

where ρ is the density of the fluid (units of mass per volume), v is the specific discharge (not the pore velocity — with units of length per time), d_{30} is a representative grain diameter for the porous medium (often taken as the 30% passing size from a grain size analysis using sieves), and μ is the dynamic viscosity of the fluid.

Derivation

Assuming stationary, creeping, incompressible flow, the Navier-Stokes equation becomes

$$\mu\nabla^2 u_i + \rho g i - 0$$

where μ is the viscosity, u_i is the velocity in the i direction, g_i is the gravity component in the i direction and p is the pressure. Assuming the viscous resisting force is proportional to the velocity, and opposite in direction, we may write

$$-\frac{\mu\phi}{\kappa_i}u_i + \rho g i - \partial i P = 0$$

where φ is the porosity.This gives the velocity

$$u_i = -\frac{\kappa_i}{\phi\mu}(\partial i P - \rho g i)$$

which gives Darcy's law

$$q_i = -\frac{\kappa_i}{\mu}(\partial i P - \rho g i)$$

ADDITIONAL FORMS OF DARCY'S LAW

Time Derivative of Flux

For very short time scales, a time derivative of flux may be added to Darcy's law, which results in valid solutions at very small times (in heat transfer, this is called the modified form of Fourier's law),

$$\tau\frac{\partial q}{\partial t} + q = -K\nabla h,$$

where τ is a very small time constant which causes this equation to reduce to the normal form of Darcy's law at "normal" times (> nanoseconds). The main reason for doing this is that the regular groundwater flow equation (diffusion equation) leads to singularities at constant head boundaries at very small times. This form is more mathematically rigorous, but leads to a hyperbolic groundwater flow equation, which is more difficult to solve and is only useful at very small times, typically out of the realm of practical use.

Brinkman Term

Another extension to the traditional form of Darcy's law is the Brinkman term, which is used to account for transitional flow between boundaries,

$$\beta\nabla^2 q + q = -K\nabla h,$$

where β is an effective viscosity term. This correction term accounts for flow through medium where the grains of the media are porous themselves, but is difficult to use, and is typically neglected.

Multiphase Flow

For multiphase flow, an approximation is to use Darcy's law for each phase, with permeability replaced by phase permeability, which is the permeability of the rock multiplied with relative permeability. This approximation is valid if the interfaces between the fluids remain static, which is not true in general, but it is still a reasonable model under steady-state conditions.

Assuming that the flow of a phase in the presence of another phase can be viewed as single phase flow through a reduced pore network, we can add the subscript *i* for each phase to Darcy's law above written for Darcy flux, and obtain for each phase in multiphase flow

$$q_i = -\frac{\kappa_i}{\mu_i} \nabla P_i,$$

where κ_i is the *phase permeability* for phase *i*. From this we also define relative permeability κ_{ri} for phase *i* as

$$\kappa_{ri} = \kappa_i / \kappa$$

where κ is the permeability for the porous medium, as in Darcy's law.

Forchheimer Equation for Non-Darcy Flow

For a sufficiently high flow velocity, the flow is nonlinear, and Dupuit and Forchheimer has proposed to generalize the flow equation to

$$\nabla P = \frac{\mu}{\kappa} V + \beta \rho V^2$$

where *V* is the flow velocity and β is a factor to be experimentally deduced.

In Membrane Operations

In pressure-driven membrane operations, Darcy's law is often used in the form,

$$J = \frac{\Delta P - \Delta \Pi}{\mu(R_f + R_m)}$$

where,

- *J* is the volumetric flux ($m.s^{-1}$),
- ΔP is the hydraulic pressure difference between the feed and permeate sides of the membrane (*Pa*),

- Δ Ð is the osmotic pressure difference between the feed and permeate sides of the membrane (*Pa*),
- μ is the dynamic viscosity (*Pa.s*),
- R_f is the fouling resistance (m^{-1}), and
- R_m is the membrane resistance (m^{-1}).

GROUNDWATER FLOW EQUATION

Used in hydrogeology, the groundwater flow equation is the mathematical relationship which is used to describe the flow of groundwater through an aquifer. The transient flow of groundwater is described by a form of the diffusion equation, similar to that used in heat transfer to describe the flow of heat in a solid (heat conduction). The steady-state flow of groundwater is described by a form of the Laplace equation, which is a form of potential flow and has analogs in numerous fields.

The groundwater flow equation is often derived for a small representative elemental volume (REV), where the properties of the medium are assumed to be effectively constant. A mass balance is done on the water flowing in and out of this small volume, the flux terms in the relationship being expressed in terms of head by using the constituative equation called Darcy's law.

MASS BALANCE

A mass balance must be performed, along with Darcy's law, to arrive at the transient groundwater flow equation. This balance is analogous to the energy balance used in heat transfer to arrive at the heat equation. It is simply a statement of accounting, that for a given control volume, aside from sources or sinks, mass cannot be created or destroyed. The conservation of mass states that for a given increment of time (Ät) the difference between the mass flowing in across the boundaries, the mass flowing out across the boundaries, and the sources within the volume, is the change in storage.

$$\frac{\Delta M_{stor}}{\Delta t} = \frac{\Delta M_{in}}{\Delta t} - \frac{\Delta M_{out}}{\Delta t} - \frac{\Delta M_{gen}}{\Delta t}$$

Diffusion Equation (Transient Flow)

Mass can be represented as density times volume, and under most conditions, water can be considered incompressible (density does not depend on pressure). The mass fluxes across the boundaries then become volume fluxes (as are found in Darcy's law). Using Taylor series to represent the in and out flux terms across the boundaries of the control volume, and using the divergence theorem to turn the flux across the boundary into a flux over the entire volume, the final form of the groundwater flow equation (in differential form) is:

$$S_S \frac{\partial h}{\partial t} = -\nabla \cdot q - G$$

This is known in other fields as the diffusion equation or heat equation, it is a parabolic partial differential equation (PDE). This mathematical statement indicates that the change in hydraulic head with time (left hand side) equals the negative divergence of the flux (q) and the source terms (G). This equation has both head and flux as unknowns, but Darcy's law relates flux to hydraulic heads, so substituting it in for the flux (q) leads to

$$S_S \frac{\partial h}{\partial t} = -(-K\nabla h) - G$$

Now if hydraulic conductivity (k) is spatially uniform and isotropic (rather than a tensor), it can be taken out of the spatial derivative, simplifying them to the Laplacian, this makes the equation

$$S_S \frac{\partial h}{\partial t} = -\kappa \nabla^2 h - G.$$

Dividing through by the specific storage (S_s), puts hydraulic diffusivity α (= k/S_s or equivalently, $\alpha = T/S$) on the right hand side. The hydraulic diffusivity is proportional to the speed at which a finite pressure pulse will propagate through the system (large values of α lead to fast propagation of signals). The groundwater flow equation then becomes

$$\frac{\partial h}{\partial t} = \alpha \nabla^2 h - G.$$

Where the sink/source term, G, now has the same units, but a different definition.

RECTANGULAR CARTESIAN COORDINATES

Especially when using rectangular grid finite-difference models, we deal with Cartesian coordinates. In these coordinates the general Laplacian operator becomes (for three-dimensional flow) specifically

$$\frac{\partial h}{\partial t} = \alpha \left[\frac{\partial^2 h}{\partial x^2} + \frac{\partial^2 h}{\partial y^2} + \frac{\partial^2 h}{\partial z^2} \right] - G.$$

As an aside, MODFLOW is actually a "quasi 3D" simulation; it only deals with the vertically averaged T and S, rather than k and S_s. In the PDE solved by MODFLOW there is no vertical (z-direction) derivative, flow is calculated between 2D horizontal layers using the concept of leakage.

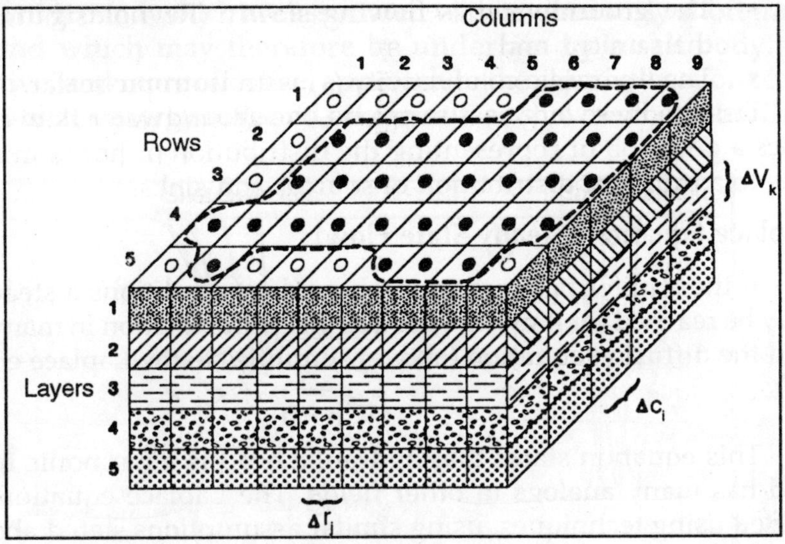

Fig. Three-Dimensional Finite Difference Grid Used in MODFLOW

Circular Cylindrical Coordinates

Another useful coordinate system is 3D cylindrical coordinates (typically where a pumping well is a line source located at the origin — parallel to the *z* axis — causing converging radial flow). Under these conditions the above equation becomes (*r* being radial distance and *è* being angle),

$$\frac{\partial h}{\partial t} = \alpha\left[\frac{\partial^2 h}{\partial \gamma^2} + \frac{1}{\gamma}\frac{\partial h}{\partial \gamma} + \frac{1}{\gamma^2}\frac{\partial^2 h}{\partial \theta^2} + \frac{\partial^2 h}{\partial z^2}\right] - G.$$

Assumptions

This equation represents flow to a pumping well (a sink of strength G), located at the origin. Both this equation and the Cartesian version above are the fundamental equation in groundwater flow, but to arrive at this point requires considerable simplification. Some of the main assumptions which went into both these equations are:

- The aquifer material is incompressible (no change in matrix due to changes in pressure — aka subsidence),
- The water is of constant density (incompressible),
- any external loads on the aquifer (e.g., overburden, atmospheric pressure) are constant,

- For the 1D radial problem the pumping well is fully penetrating a non-leaky aquifer,
- The groundwater is flowing slowly (Reynolds number less than unity), and
- The hydraulic conductivity (k) is an isotropic scalar.

Despite these large assumptions, the groundwater flow equation does a good job of representing the distribution of heads in aquifers due to a transient distribution of sources and sinks.

Laplace Equation (Steady-State Flow)

If the aquifer has recharging boundary conditions a steady-state may be reached (or it may be used as an approximation in many cases), and the diffusion equation (above) simplifies to the Laplace equation.

$$\nabla^2 h = 0$$

This equation states that hydraulic head is a harmonic function, and has many analogs in other fields. The Laplace equation can be solved using techniques, using similar assumptions stated above, but with the additional requirements of a steady-state flow field.

A common method for solution of this equations in civil engineering and soil mechanics is to use the graphical technique of drawing flownets; where contour lines of hydraulic head and the stream function make a curvilinear grid, allowing complex geometries to be solved approximately.

Steady-state flow to a pumping well (which never truly occurs, but is sometimes a useful approximation) is commonly called the Thiem solution.

Two-Dimensional Groundwater Flow

The above groundwater flow equations are valid for three dimensional flow. In unconfined aquifers, the solution to the 3D form of the equation is complicated by the presence of a free surface water table boundary condition: in addition to solving for the spatial distribution of heads, the location of this surface is also an unknown. This is a non-linear problem, even though the governing equation is linear.

An alternative formulation of the groundwater flow equation may be obtained by invoking the Dupuit assumption (or Dupuit-Forcheimer assumption), where it is assumed that heads do not vary in the vertical direction (i.e. $\partial h / \partial z = 0$,). A horizontal water balance is applied to a long vertical column with area äxäy extending from the aquifer base to the unsaturated surface. This distance is referred to as the saturated thickness, b. In a confined aquifer, the saturated thickness is determined by the height of the aquifer, H, and the pressure head is non-zero everywhere. In an unconfined aquifer, the saturated thickness is defined as the vertical distance between the water table surface and the aquifer base. If $\partial h / \partial z = 0$, and the aquifer base is at the zero datum, then the unconfined saturated thickness is equal to the head, i.e., b=h.

Assuming both the hydraulic conductivity and the horizontal components of flow are uniform along the entire saturated thickness of the aquifer (i.e., $\partial q_x / \partial z = 0 \text{and} \partial \kappa / \partial z = 0$), we can express Darcy's law in terms of integrated discharges, Q_x and Q_y:

$$Qx = \int_0^b qxdz = -\kappa b \frac{\partial h}{\partial x}$$

$$Qx = \int_0^b q_y dz = -\kappa b \frac{\partial h}{\partial y}$$

Inserting these into our mass balance expression, we obtain the general 2D governing equation for incompressible saturated groundwater flow:

$$\frac{\partial nb}{\partial t} - \nabla \cdot (\kappa b \nabla h) + N.$$

Where *n* is the aquifer porosity. The source term, *N* (length per time), represents the addition of water in the vertical direction (e.g., recharge). By incorporating the correct definitions for saturated thickness, specific storage, and specific yield, we can transform this into two unique governing equations for confined and unconfined conditions:

$$S \frac{\partial h}{\partial t} = \nabla \cdot (\kappa H \nabla h) + N.$$

(confined), where $S=S_s b$ is the aquifer storativity and

$$\int_0^b qxdz = -\kappa b S_y \frac{\partial h}{\partial t} = \nabla \cdot (\kappa H \nabla h) + N.$$

(unconfined), where S_y is the specific yield of the aquifer.

Note that the partial differential equation in the unconfined case is non-linear, whereas it is linear in the confined case. For unconfined steady-state flow, this non-linearity may be removed by expressing the PDE in terms of the head squared:

$$\nabla \cdot \left(\kappa \nabla h^2 \right) = -2N.$$

Or, for homogeneous aquifers,

$$\nabla^2 h^2 = -\frac{2N}{\kappa}.$$

This formulation allows us to apply standard methods for solving linear PDEs in the case of unconfined flow. For heterogeneous aquifers with no recharge, Potential flow methods may be applied for mixed confined/unconfined cases.

WATER CONTENT

Water content or moisture content is the quantity of water contained in a material, such as soil (called soil moisture), rock, ceramics, or wood on a volumetric or gravimetric basis. The property is used in a wide range of scientific and technical areas, and is expressed as a ratio, which can range from 0 (completely dry) to the value of the materials' porosity at saturation.

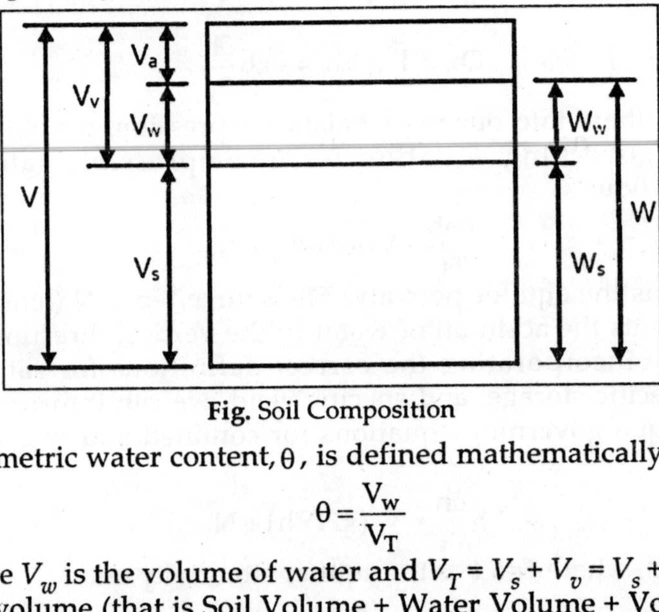

Fig. Soil Composition

Volumetric water content, θ, is defined mathematically as:

$$\theta = \frac{V_W}{V_T}$$

where V_w is the volume of water and $V_T = V_s + V_v = V_s + V_w + V_a$ is the total volume (that is Soil Volume + Water Volume + Void Space). Water content may also be based on its mass or weight, thus the gravimetric water content is defined as:
where m_w is the mass of water and m_b (or m_s for soil) is the bulk material mass. To convert gravimetric water content to volumetric water, multiply the gravimetric water content by the bulk specific gravity of the material.

OTHER DEFINITIONS

Degree of Saturation

In soil mechanics and petroleum engineering, the term water saturation or degree of saturation, S_w is used, defined as

$$S_w = \frac{V_w}{V_v} = \frac{V_w}{V_v \phi} = \frac{\theta}{\phi}$$

where $\varphi = V_v / V_T$ is the porosity and V_v is the volume of void or pore space.

Values of S_w can range from 0 (dry) to 1 (saturated). In reality, S_w never reaches 0 or 1 - these are idealizations for engineering use.

Normalized Volumetric Water Content

The normalized water content, (also called effective saturation or S_e) is a dimensionless value defined by van Genuchten as:

$$\Theta = \frac{\theta - \theta_r}{\theta_s - \theta_r}$$

where θ is the volumetric water content; θ_r is the residual water content, defined as the water content for which the gradient $d\theta / dh$ becomes zero; and, θ_s is the saturated water content, which is equivalent to porosity, φ.

MEASUREMENT

Direct methods

Volumetric water content can be directly measured using a known volume of the material, and a drying oven. Volumetric water content, θ, is calculated using:

$$\theta = \frac{m_{wet} - m_{dry}}{\rho_w \cdot V_b}$$

where

m_{wet} and m_{dry} are the masses of the sample before and after drying in the oven;

ρ_w is the density of water; and

V_b is the volume of the sample before drying the sample.

For materials that change in volume with water content, such as coal, the water content, u, is expressed in terms of the mass of water per unit mass of the moist specimen:

$$u = \frac{m_{wet} - m_{dry}}{m_{wet}}$$

However, geotechnics requires the moisture content to be expressed as a percentage of the sample's dry weight i.e. % moisture content = u * 100

Where

$$u = \frac{m_{wet} - m_{dry}}{m_{dry}}$$

For wood, the convention is to report moisture content on oven-dry basis (i.e. generally drying sample in an oven set at 105 deg Celsius for 24 hours). In wood drying, this is an important concept.

Laboratory Methods

Other methods that determine water content of a sample include chemical titrations (for example the Karl Fischer titration), determining mass loss on heating (perhaps in the presence of an inert gas), or after freeze drying. In the food industry the Dean-Stark method is also commonly used.

From the Annual Book of ASTM (American Society for Testing and Materials) Standards, the total evaporable moisture content in Aggregate (C 566) can be calculated with the formula:

$$p = \frac{W - D}{D}$$

where p is the fraction of total evaporable moisture content of sample, W is the mass of the original sample, and D is mass of dried sample.

GEOPHYSICAL METHODS

There are several geophysical methods available that can approximate *in situ* soil water content. These methods include: time-domain reflectometry (TDR), neutron probe, frequency domain sensor, capacitance probe, electrical resistivity tomography, ground penetrating radar (GPR), and others that are sensitive to the physical properties of water. Geophysical sensors are often used to monitor soil moisture continuously in agricultural and scientific applications.

Satellite Remote Sensing Method

Satellite microwave remote sensing is used to estimate soil moisture based on the large contrast between the dielectric properties of wet and dry soil. The data from microwave remote sensing satellite such as: WindSat, AMSR-E, RADARSAT, ERS-1-2 are used to estimate surface soil moisture.

Classification and Uses

Moisture may be present as adsorbed moisture at internal surfaces and as capillary condensed water in small pores. At low relative humidities, moisture consists mainly of adsorbed water. At higher relative humidities, liquid water becomes more and more important, depending on the pore size. In wood-based materials, however, almost all water is adsorbed at humidities below 98% RH.

In biological applications there can also be a distinction between physisorbed water and "free" water — the physisorbed water being that closely associated with and relatively difficult to remove from a biological material. The method used to determine water content may affect whether water present in this form is accounted for. For a better indication of "free" and "bound" water, the water activity of a material should be considered. Water molecules may also be present in materials closely associated with individual molecules, as "water of crystallization", or as water molecules which are static components of protein structure.

Earth and Agricultural Sciences

In soil science, hydrology and agricultural sciences, water content has an important role for groundwater recharge, agriculture, and soil chemistry. Many recent scientific research efforts have aimed toward a predictive-understanding of water content over space and time. Observations have revealed generally that spatial variance in water content tends to increase as overall wetness increases in semiarid regions, to decrease as overall wetness increases in humid regions, and to peak under intermediate wetness conditions in temperature regions.

There are four standard water contents that are routinely measured and used, which are described in the following table:

Name	Notation	Suction pressure (J/kg or KPa)	Typical water content(vol/vol)	Description
Saturated Water content	θ_s	0	0.2-0.5	Fully saturated water,equivalent to effective porosity
Field capacity	θ_{fc}	-33	0.1-0.35	Soil moisture after 2-3 days after a rain or irrigation
Permanent wilting point	θ_{pwp} or θ_{wp}	-1500	0.001-0.25	Minimum soil moisture at which a plant wilts
Residual water content	θ_r	-00	0.001-0.1	Remanining water at high tension

And lastly the available water content, θ_a, which is equivalent to:

θ_a a″ θ_{fc} ″ θ_{pwp}
which can range between 0.1 in gravel and 0.3 in peat.

Agriculture

When a soil gets too dry, plant transpiration drops because the water is becoming increasingly bound to the soil particles by suction. Below the wilting point plants are no longer able to extract water. At this point they wilt and cease transpiring altogether. Conditions where soil is too dry to maintain reliable plant growth is referred to as agricultural drought, and is a particular focus of irrigation management. Such conditions are common in arid and semi-arid environments.

Some agriculture professionals are beginning to use environmental measurements such as soil moisture to schedule irrigation. This method is referred to as "Smart Irrigation."

Groundwater

In saturated groundwater aquifers, all available pore spaces are filled with water (volumetric water content = porosity). Above a capillary fringe, pore spaces have air in them too.

Most soils have a water content less than porosity, which is the definition of unsaturated conditions, and they make up the subject of vadose zone hydrogeology. The capillary fringe of the water table is the dividing line between saturated and unsaturated conditions. Water content in the capillary fringe decreases with increasing distance above the phreatic surface.

One of the main complications which arises in studying the vadose zone, is the fact that the unsaturated hydraulic conductivity is a function of the water content of the material. As a material dries out, the connected wet pathways through the media become smaller, the hydraulic conductivity decreasing with lower water content in a very non-linear fashion.

A water retention curve is the relationship between volumetric water content and the water potential of the porous medium. It is characteristic for different types of porous medium. Due to hysteresis, different wetting and drying curves may be distinguished.

Chapter 11

Greenhouse Gases

THE FOOD SYSTEM

The food system encompasses all the activities associated with providing food to consumers. These include food production (e.g., onsite clearing, cultivation, and harvesting, and offsite production of fertilizer and other agricultural inputs) and post-harvest activities (e.g., food processing, transport, cooking).

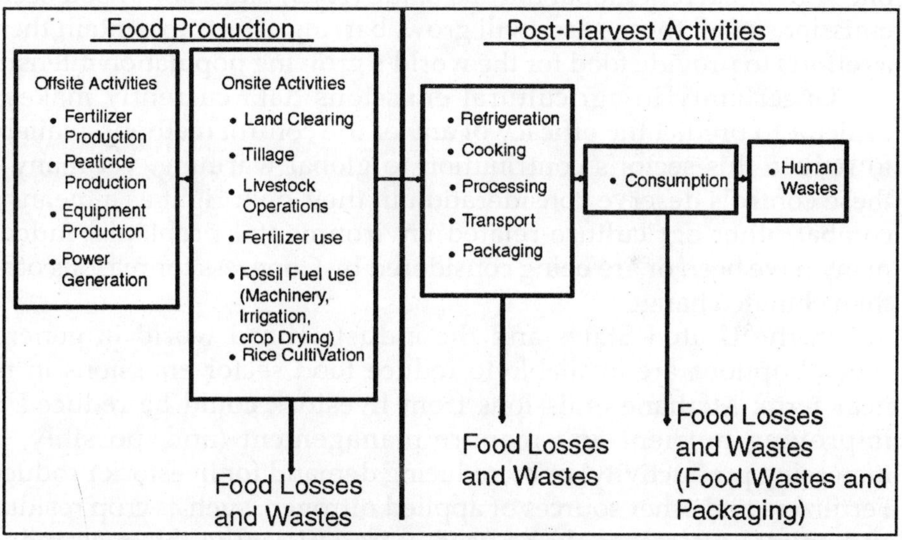

Fig.The Food System

Congress has become increasingly aware of the environmental impacts of the food system and has begun to address some of these through legislation. In addition to previously identified environmental impacts of agriculture (pollution of surface and groundwater by nitrate and pesticides, soil erosion, depletion of aquifers to meet irrigation needs, loss of natural habitat), the food system is now also recognized as a potential contributor to global climate change.

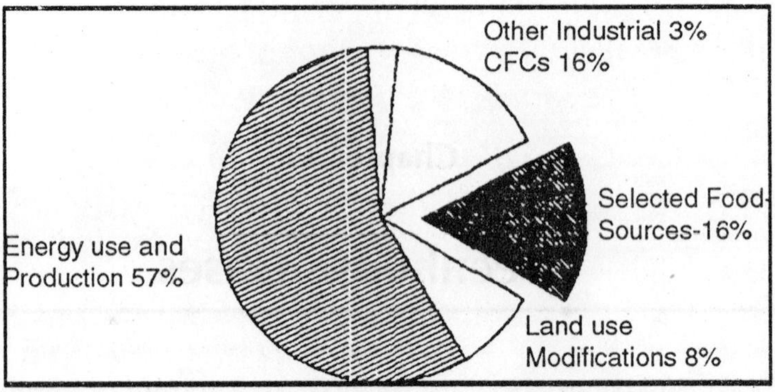

Fig. Contribution of Selected Food-Related

Figure shows the global food system's contribution to global warming in the 1980s. Although estimates are uncertain, the food sector may account for one-third of global methane (CH_4) emissions; one-fifth of net global carbon dioxide (CO_2) emissions; up to 15 per cent of chlorofluorocarbon (CFC) emissions; and anywhere from one-tenth to one-fifth of current global nitrous oxide (N_2O) emissions. Food sector emissions of all these gases will grow, barring efforts to contain them, as efforts to provide food for the world's growing population intensify.

Uncertainty in agricultural emissions data currently makes it difficult to predict the efficacy of any of the control methods available to reduce this sector's contribution to global warming, yet many of these controls deserve consideration in their own right as a means to combat other agriculture-related environmental problems. Indeed, many have been or are being considered by Congress for reasons other than climate change.

In the United States and the industrialized world in general. several options are available to reduce food sector emissions in the near term. Methane emissions from livestock could be reduced by improving nutrient and manure management (and, possibly, by increasing productivity) or by reducing demand for livestock products. Fertilizers and other sources of applied nitrogen, such as crop residues and animal wastes, could be used more efficiently; this may reduce N_2O emissions, as well as surface and groundwater contamination, and would help conserve soil organic matter.

Nitrogen fertilizer manufacturing and onsite farm machinery and cultivation practices could be more energy efficient; while reductions in CO_2 emissions would be relatively small, other benefits such as decreased local air pollution from fossil fuel combustion would accrue. Land transformations that help remove carbon from the atmosphere (such as converting cropland to forest land) could be encouraged, while those that increase CO_2 emissions could be discouraged. In food refrigeration, emissions can be curbed by preventing the release of CFCs from existing refrigerators and eliminating their future use; and by improving energy efficiencies. Further CO_2 reductions could be achieved by designing stoves and ovens that use energy more efficiently and by increasing fuel efficiency in vehicles used in food transport.

In developing countries, slowing deforestation, maintaining or increasing crop yields, and reducing emissions associated with cooking can be more effective, in the short term, than changing current patterns of fossil fuel use.

Alternatives to clearing of tropical forests include increased use of agroforestry and "sustainable" agriculture, and decreased subsidies for cattle ranching on lands that cannot support livestock for more than a few years.

Crop yields may be improved on existing agricultural land with increased agricultural inputs (e.g., fertilizers, irrigation, etc.), but this may increase greenhouse gas emissions per acre. Opportunities exist to reduce methane emissions from livestock through technology transfer to developing countries, but the relevant technologies may not be readily applicable in most developing countries. In general, direct U.S. Government influence in these areas tends to be through the U.S. Agency for International Development (A.I.D.), research programs, multinational corporations, and participation in multilateral lending institutions.

EMISSIONS FROM THE FOOD SYSTEM

Activities in the food system affect the flows of many substances to and from the atmosphere, including small particles (aerosols) and numerous trace gases.

In some developing countries, food production (i.e., activities up to and including harvest) is the dominant source of greenhouse gases, primarily because of CO_2 emitted through land transformations (e.g., land clearing and field burning) and CH_4 emitted through rice cultivation. Although global data on emissions from post-harvest activities are poor, these activities are likely the most important source of emissions in *industrialized countries.*

Food Production

Greenhouse gas sources from food production activities include:

- Flooded rice fields, which are significant sources of CH_4, particularly in the developing countries of Asia;

- Livestock, a significant source of CH_4 in many industrialized and developing countries, through direct emissions from animals as well as their manure;
- Nitrogenous fertilizers, the use of which results in N_2O;
- Large-scale land transformations (e.g., clearing tropical forests) in many developing countries; and, to a lesser extent, land use changes in industrialized countries (e.g., urbanization), both of which result in CO_2 emissions; and
- Burning of vegetation to clear and/or prepare land for agriculture (especially in developing countries), which adds to atmospheric concentrations of several gases, including CO_2, CH_4, N_2O, and carbon monoxide.

In general, emissions from food production and their contributions to climate change are difficult to quantify because of the complex interactions of biological and chemical processes in soils, water, plants, animals, and the atmosphere, and because studies are lacking for many topics and areas.

Nevertheless, some approximations and projections are possible. For example, EPA estimates that if current agricultural practices continue, CH_4 emissions from rice will increase by about 35 per cent by 2025, those from livestock will rise by about 65 per cent, and N_2O emissions from fertilizers could more than double.

Methane (CH_4)

Methane is produced when bacteria decompose organic material in oxygen-deficient (i.e., anaerobic) environments, such as sediments at the bottom of flooded or rainfed rice paddies, landfills, and the digestive tract of ruminant animals and termites; CH_4 is also emitted as a byproduct when wood and other biomass are burned. Emissions from these and other CH_4 sources are very poorly characterized, partly because they vary enormously on geographic scales and over time. Available evidence indicates that about one-third of total global CH_4 emissions comes from the food sector; this represents about 60 per cent of total emissions from anthropogenic sources, or roughly 10 per cent of global warming in the 1980s.

Rice Paddies

Rice provided about one-fourth of the world's cereal grains in 1985. China and India produced well over half of the total, the United States only 1 per cent. Global rice production increased by nearly 200 per cent between the Late 1940s and 1985, while acreage grew by 43 per cent.

Estimates of CH_4 released from rice paddies in the mid-1980s range from roughly 25 to 170 million metric tons, or about 15 to 30 per cent of world methane emissions from anthropogenic sources.

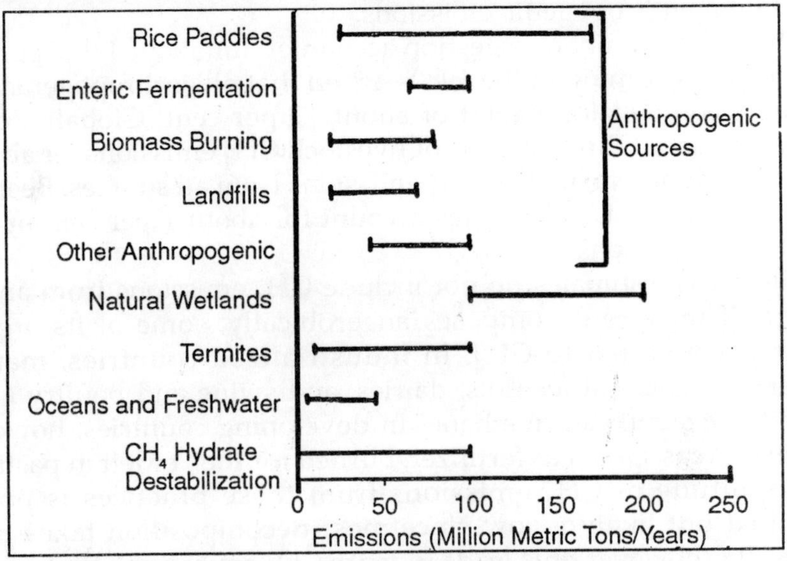

Fig. Estimated Global Emissions of Methane

Rice paddies thus accounted for about 3 to 6 per cent of the contribution to global warming in the 1980s. These emissions estimates are based on a few studies conducted in temperate regions (California, Italy, and Spain), but because CH_4 emissions are likely to be highly site-specific, the extrapolation of these data to other regions (including tropical areas, where most of the world's rice is grown) is problematic. However, new data from Japan, China, and the Philippines are now becoming available.

Recent field tests, for example, have shown that during the growing season, rice fields in China emit up to 10 times more CH_4 per hectare per hour than rice fields in Europe and the United States. If new data from China are representative of conditions in the Far East (over 90 per cent of the world's rice is produced in Asia), then CH_4 emissions from rice cultivation are higher than the estimates presented above, although it is difficult to estimate by how much.

Ruminants

Much of the world's livestock consists of ruminant animals—sheep, goats, camels, cattle, and buffalo. One of the unique features of ruminants is their four-chambered stomach, including one chamber called the rumen in which bacteria break down food and generate CH_4

as a byproduct. Ruminants emitted an estimated 65 to 100 million metric tons of CH_4 per year in the mid-1980s, perhaps 10 to 20 per cent of global CH_4 emissions from all sources, or about 20 to 40 per cent of total anthropogenic emissions.

Therefore, ruminant digestion accounted for about 4 to 7 per cent of the global warming in the 1980s, given that all anthropogenic CH_4 sources accounted for a total of about 19 per cent. Globally, cattle account for about three-fourths of livestock CH_4 emissions, or about 7 to 15 per cent of total global CH_4 emissions from all sources. Beef and dairy cattle in the United States account for about 1 per cent of total global CH_4 emissions.

The above estimates do not include CH_4 emissions from animal manure. If manure decomposes anaerobically, some of its organic matter is converted to CH_4. In industrialized countries, manure handling practices at feedlots, dairies, and swine and poultry farms may release significant methane. In developing countries, however, most manure is spread as fertilizer, burned for fuel, or left in pastures; the magnitude of CH_4 emissions from these practices is poorly quantified but likely is low since most decomposition takes place aerobically (however, this leads to more CO_2 emissions). Preliminary estimates suggest global CH_4 emissions from manure are on the order of about 20 to 40 million metric tons per year.

Biomass Burning

Burning vegetation contributes 20 to 80 million metric tons of methane per year, or roughly 7 to 8 per cent of global emissions from anthropogenic sources and natural sources such as lightning-induced forest fires. This portion may be higher if recent data indicating higher deforestation rates are correct. Substantial but unquantified CH_4 emissions result from fires ignited deliberately—forests burned to produce rangeland or cropland; grasslands burned to enhance forage; and crop residues burned to return nutrients to the soil. These emissions are heavily concentrated in tropical countries, where large areas of savanna and forest are burned or cleared each year for agriculture.

NITROUS OXIDE (N_2O)

Agriculture introduces nitrogenous compounds to the environment in the form of commercial fertilizers, legumes, and crop residues. N_2O emissions from soil and water occur through nitrification and denitrification of these compounds and also result when vegetation is cleared through burning.

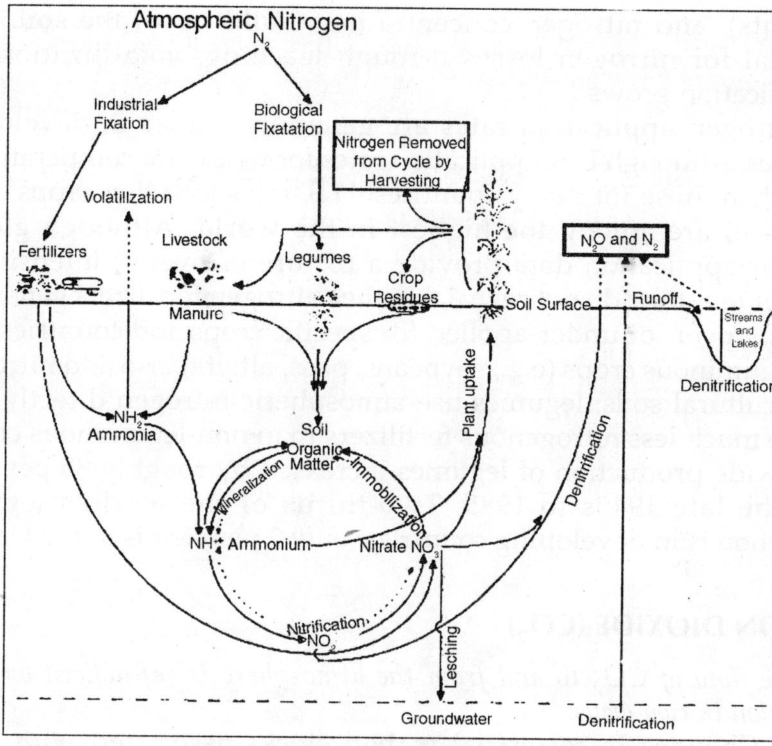

Fig. Nitrogen Cycle

The magnitude of N_2O emissions from terrestrial and aquatic sources is very poorly characterized. Based on annual increases in N_2O atmospheric concentrations, the Intergovernmental Panel on Climate Change estimates that total global N_2O emissions *should* be around 10 to 17.5 million metric tons per year; however, only 4.4 to 10 million metric tons can be accounted for from known sources.

N_2O emissions associated with fertilizer use are not well understood and probably vary with factors such as fertilizer type, method of application, and amount applied. Worldwide nitrogen fertilizer consumption is about 74 million metric tons per year. China accounted for nearly 20 per cent of this consumption in 1987, the United States for about 13 per cent. Current fertilizer-derived emissions are on the order of 0.01 to 2.2 million metric tons per year, about 0.2 to 20 per cent of global emissions from all sources, or about 10 to 80 per cent of all anthropogenic emissions.

The general relationship between nitrogen application rate and maize yields illustrates that yields are highest at a certain optimal nitrogen application rate; further additions result in either stable or even lower yields (because nitrogen is no longer used as effectively by plants), and nitrogen concentrations build up in the soil. The potential for nitrogen losses through leaching, volatilization, or denitrification grows.

Nitrogen application rates are generally higher in developed countries. Although U.S. application rates for wheat are comparatively lower than those for many countries, U.S. rates for other crops (e.g., corn, rice) are among the highest in the world. Although global fertilizer application data provide a picture of overall intensity of fertilizer use, they do not reveal whether nitrogen fertilizers generally are being over- or under-applied for specific crops and countries.

Leguminous crops (e.g., soybeans, peas, alfalfa) also add nitrogen to agricultural soils; legumes use atmospheric nitrogen directly and require much less nitrogenous fertilizers than non-leguminous crops. Worldwide production of legumes increased by roughly 85 per cent from the late 1940s to 1985. Two-thirds of the world's legume production is in developing countries; only 2 per cent is in the United States.

CARBON DIOXIDE (CO_2)

The flow of CO_2 to and from the atmosphere is influenced by food production in two ways:

- Changes in terrestrial carbon stocks associated with land transformations, and
- Emissions from fossil fuel use.

Land Use Changes and Terrestrial Carbon

Land transformations have characterized the entire 10,000-year history of agricultural development and continue on a large scale today. Although concern over deforestation now focuses on tropical areas, many temperate forests have also been cleared at least once during the last few hundred years. CO_2 is emitted in this process, and also when grasslands and savannas are burned to enhance grazing conditions and when carbon contained in soil organic matter is carried off by erosion or converted into CO_2 by microorganisms. Urbanization claims additional forest and agricultural land each year. These land transformations greatly affect how carbon is distributed in organisms, soils, and sediments, and how it flows to and from the atmosphere. The net result has increased atmospheric carbon concentrations, mostly as CO_2 but also in the form of other carbon compounds such as methane.

Today, up to one-fifth of net global CO_2 emissions may be attributed to clearing and burning tropical forests for food production.

Additional CO_2 emissions result from burning savannas and agricultural wastes, and using biomass fuels for cooking. Therefore, these activities might have accounted for roughly 8 to 10 per cent of the global warming in the 1980s. As population and economic pressures increase, the rate of deforestation could accelerate. CO_2 emissions from soil erosion may account for about 1 to 2 per cent of global emissions, but data on this pathway are sparse and very uncertain.

Fossil Fuel Combustion

Most food-related CO_2 emissions in the U.S. occur in the post-harvest phase. However, fossil fuels are also used for food production, for example to drive farm machinery such as tractors and irrigation pumps and to produce, offsite, inputs such as fertilizers and pesticides. Collectively, these uses account for a relatively small share of world fossil fuel use and for about 2 per cent of U.S. CO_2 emissions.

Onsite Fossil Fuel Use: Global data are relatively poor but suggest that farms released perhaps 76 million metric tons of carbon annually during the mid-1980s through onsite fossil fuel use (not including emissions from fertilizer and pesticide manufacture). This represents about 1 per cent of global carbon emissions from fossil fuel use and accounted for only about one-half per cent of global warming in the 1980s. Similarly, CO_2 emissions from fossil fuel use on U.S. farms during this period (about 14 million metric tons per year) represented about 1 per cent of U.S. CO_2 emissions from fossil fuels. However, both total and per-hectare energy use in the United States have declined sharply since the mid-1970s.

Offsite Nitrogen Fertilizer Manufacturing: Another 1 per cent of world CO_2 emissions from fossil fuel use, or about 0.5 per cent of the global warming in the 1980s, results from commercial fertilizer manufacture. Nitrogen fertilizers account for most of these emissions because they are produced in large amounts and their manufacture (often with natural gas as a feedstock) is very energy intensive. In the United States, nitrogen fertilizer production accounted for about 0.8 per cent of U.S. CO_2 emissions from fossil fuel use. World fertilizer production increased at a rate of 6.2 per cent annually from 1965 to 1985, and the outlook is for continued growth, especially in developing countries.

Post-Harvest Activities

Once harvested, some food is consumed directly by livestock and humans. The bulk, however, is processed, preserved (often though cooking), stored, and distributed with a certain amount of loss and waste during transport. The major emissions from these post-harvest activities include:

- CFCs from refrigeration;
- CO_2 from fossil fuels used in food processing, refrigeration, transport, and cooking, and from biomass fuels used for cooking; and
- CH_4 from decomposition of food-related wastes (including packaging).

Some of these post-harvest activities (e.g., food transport, refrigeration) fall within sectors examined elsewhere in the report but are highlighted here.

CHLOROFLUOROCARBONS (CFCS)

CFCs used in refrigeration are emitted through refrigerant leaks, intentional venting (during manufacture, disposal, or repair), and deterioration of insulation. Worldwide, about 47,000 metric tons of CFC-11 and 85,000 metric tons of CFC-12—roughly 15 per cent of total consumption—are used for food-related refrigeration.

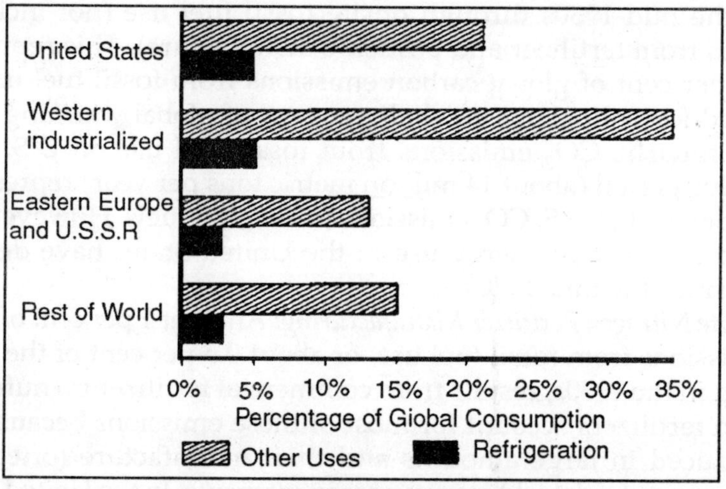

Fig. Consumption of CFC-11 Plus CFC-12 for Food-Related Redrigeration and othe Uses

CFCs used in the food system accounted for about 2 per cent of the global warming in the 1980s, assuming that all are eventually emitted to the atmosphere and that *total* CFC-11 and -12 uses accounted for about 14 per cent (143). Most CFC use in refrigeration occurs in cold storage warehouses, retail refrigeration, and refrigerated transport. Lesser, but still substantial, amounts are used in residential refrigerators and freezers.

Global use of CFCs has increased dramatically during the 60 years they have been commercially produced. From 1976 to 1986, despite mounting evidence of the ozone-depleting properties of CFCs, CFC-12 refrigerant sales increased by over 50 per cent. The United States used 41,000 metric tons of CFC-11 and CFC-12 for food refrigeration in 1985. This represents one-third of world use for food-related refrigeration and about 19 per cent of total U.S. use of the compounds.

CFC use is expected to rise rapidly in developing countries. With the promise of funding from the industrialized world, however, key countries such as China and India are expected to join the Montreal Protocol and pledge to reduce and eventually phase out CFC use.

CARBON DIOXIDE (CO_2)

In developing countries, residential cooking is by far the most significant source of post-harvest CO_2 emissions. In industrialized countries, the most significant post-harvest activities are processing in the food industry, residential cooking, and refrigeration (i.e., powerplant emissions attributed to the energy needs of refrigeration). In the United States, energy used for residential and supermarket refrigeration, residential cooking, and food processing and packaging accounted for about 5 per cent of U.S. CO_2 emissions from fossil fuel use, or roughly 1 per cent of global emissions from fossil fuels. Additional CO_2 emissions arise from other activities (e.g., energy use in food wholesaling, restaurants, and food transport).

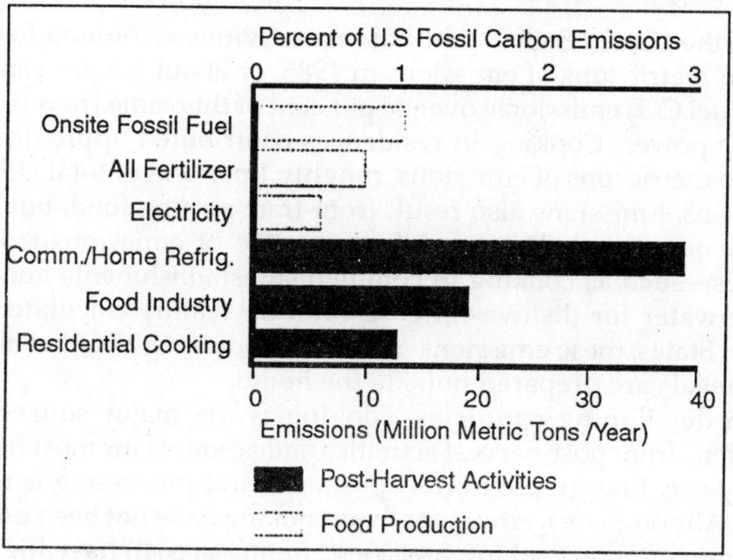

Fig. CO_2Emissions From Selected Fossil Fuel Uses

In general, more fossil fuel is used in industrialized countries for post-harvest activities than during food production. Even so, CO_2 emissions from post-harvest activities are relatively small compared with those from the energy, building, transportation, and manufacturing sectors.

Food Refrigeration-

Accurate estimates of worldwide CO_2 emissions from refrigeration are not readily available, although estimates for specific countries suggest they are important within the food sector but small relative to other sectors. In the United States, energy use for household and supermarket refrigeration accounted for about 3 per cent of total U.S. CO_2 emissions in 1985. Of this, two-thirds (24 million metric tons) can be attributed to domestic refrigeration, and about one-third to refrigeration at supermarkets. As refrigerators have "saturated" the market in industrialized countries and average energy efficiencies have improved, growth in energy use for refrigeration in these countries has slowed. By contrast, refrigeration in China accounts for only a small fraction of national energy use (only 0.4 per cent of which goes to generate all electricity), but this is rapidly changing. Between 1979 and 1987, for example, China's production of refrigerators increased by a factor of 125, and many of the several hundred million households that still do not have one may acquire one in the next decade.

Processing, Transportation, and Cooking

In the United States, post-harvest activities accounted for over 19 million metric tons of emissions in 1985, or about 1.5 per cent of U.S. fossil fuel CO_2 emissions; over 40 per cent of this came from purchased electric power. Cooking in residences contributed approximately 12 million metric tons of emissions, roughly 1 per cent of total U.S. carbon emissions. Emissions also result from transporting food, but they are poorly quantified. The overall magnitude of emissions from other sources—such as cooking in commercial establishments and heating of hot water for dishwashing—cannot be readily calculated. In the United States, these emissions could be important given that about half of all meals are prepared outside the home.

In developing countries, cooking is the major source of CO_2 emissions from post-harvest activities and accounts for most household energy use. Energy use in transportation and processing is relatively small. Although CO_2 emissions from cooking have not been quantified, the most common cooking fuels (e.g., biomass, coal) have high carbon content. Traditional biomass fuels (animal dung, crop residues, wood, etc.), which may account for as much as 15 per cent of world energy use, are used extensively for cooking and food processing (e.g., for drying). Coal also is very important in some regions. Over one-fourth of coal use in China, the world's largest coal consumer, is for domestic purposes, primarily for cooking.

METHANE (CH_4)

Following preparation and consumption of food, solid wastes (e.g., food residues, packaging) and sewage are generated. Under some disposal conditions, these wastes result in the emission of CH_4 to the atmosphere. In landfills, for example, carbon-containing compounds decompose in two stages—first aerobically, producing CO_2 emissions; then, once oxygen is used up, anaerobically (i.e., without oxygen) by methane-producing bacteria. Landfills may emit about 30 to 70 million metric tons of CH_4 annually worldwide, about one-half of what livestock emit. In the United States, about 6,000 municipal solid waste landfills were operating in 1986.

ALTERNATIVE PRACTICES

This section discusses alternative practices that could be pursued to reduce future greenhouse gas emissions from the food sector. While the overall costs and benefits of these practices are not clearly defined, collectively they could substantially reduce some emissions. Many would carry other environmental benefits as well—e.g., reducing water pollution from croplands, reducing soil erosion, conserving water supplies, preserving biological diversity, and reducing waste generation in food processing. However, tradeoffs may be associated with some alternatives. For example, some tillage practices used to conserve soils require more pesticides. Also, to reduce the pressure to open new lands to agriculture, crop yields must be increased on existing acreage, which may require greater use of fertilizer and other inputs.

Livestock

Livestock directly produce about 10 to 20 per cent of the world's CH_4 emissions through digestive processes, and indirectly produce additional emissions from anaerobic decomposition of manure. They also indirectly account for emissions of CO_2 and N_2O by virtue of the land and agricultural inputs required to sustain them.

Decrease Methane Emissions Per Unit of Output

Opportunities exist for reducing, or at least limiting, the growth rate of CH_4 emissions from livestock by increasing digestion efficiency and/or animal productivity (i.e., the amount of animal product produced per unit of feed). Emissions reductions on the order of 25 to 75 per cent per unit of product are thought to be possible, with most potential for change in developing countries (industrialized countries have already made strides in raising more productive animals). This range of possible emission reductions roughly corresponds to a 2- to 5-per cent reduction in global CH_4 emissions.

Specific options include:

- Supplement the diets of grazing animals to correct nutrient deficiencies often found in lower quality forage;
- Substitute feeds with low CH_4-producing potential for feeds with higher CH_4-producing potential in the diets of animals in confined (as opposed to free-ranging livestock):
- Develop feed additives that increase digestion efficiency and reduce methanogenesis in the rumen;
- Use growth promotants (e.g., bovine somatotropin); and
- Improve reproductive efficiency.

Enhancing animal productivity can reduce CH_4 emissions given the following assumptions: that by increasing productivity, the same amount of output could be obtained from a smaller herd size; that the rate of output (e.g., milk, beef) per animal increases faster than emissions of CH_4 per animal; and that consumer demand for these products remains relatively stable. It is also assumed that more productive cattle could be brought to market sooner, decreasing the total lifetime CH_4 emissions per animal. the assumption that increased productivity could lead to fewer livestock (and, thus, lower total CH_4 emissions) could be challenged, through. In fact, a large unmet demand for cattle products in developing countries could lead to greater livestock populations even if productivity increases.

Many productivity-enhancing/methane-reducing strategies have been used with great success in the United States and could be transferred to other countries. However, for some countries, especially in the developing world where most cattle graze in unconfined situations, it is difficult to determine how effective some of these options would be.

For example, feed additives called "ionophores," which can increase digestion efficiency, currently are feasible only for confined cattle and therefore may not be of immediate utility in developing countries. Growth hormones, such as bovine somatotropin, have been used in nondairy cattle to increase productivity per animal.

Another option, increasing the reproductive efficiency of animals,

beneficial practice for determining the availability of phosphorus, potassium, and other nutrients. Nitrogen recommendations for corn grown in the middle Atlantic and southeastern states have not been based on traditional soil test calibration methods, in contrast to other nutrients, because:

- Predicting the availability to crops of organic N has not been successful,
- The storage of soil mineral N is often too brief due to the winter and spring leaching losses of the mobile NO_3-N form that occur under our high rainfall and moderate soil temperatures.

Nitrogen analysis of animal manures, biosolids, and legumes to predict the amounts of N available for crops has been only partly successful, which accentuates the need for a soil N test.

Nitrogen behaviour in the soil is difficult to predict because N transformations in soil are very complex. Over 98% of the N in most soils is unavailable for plant uptake at any specific time because it is fixed in soil organic matter or in clay minerals. Nitrogen in organic matter (e.g., plant residue, cover crops, animal manure, biosolids) may undergo microbial transformations that convert it to a plant-available form. The end result of this process is the NO_3 form of N. The transformation rate of organic N to plant-available NO_3 in the spring increases as the soil temperature increases, and much plant-available N can be produced from organic N in Virginia, beginning in mid-May in eastern Virginia to mid-June in western Virginia.

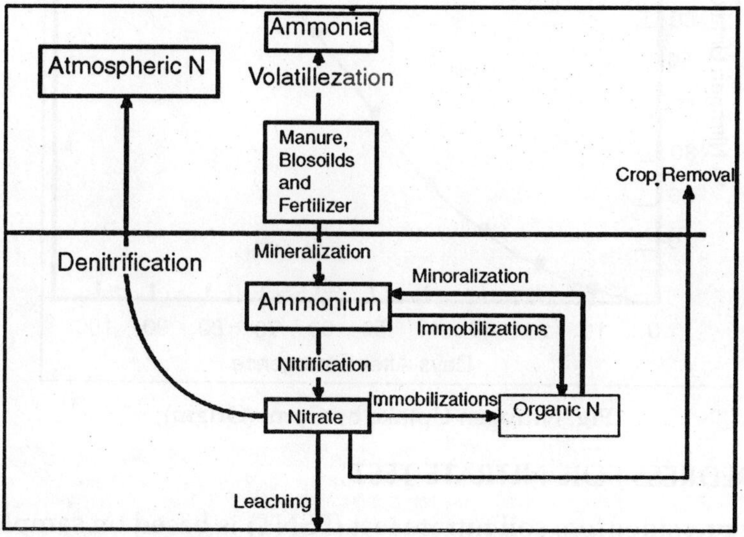

Fig. The Nitrogen Cycle in Soil. (Freelance)

The behaviour of N in the soil has several important implications for efficient N management of corn. Corn requires only small amounts of N during the first month of growth because the plants are small and root systems are not well-developed. Nitrogen applied preplant or released from organic matter during early spring can be lost by leaching during this time when the plants' N requirements are low and soil moisture is high. Therefore, only small amounts of starter N (25-30 lbs/acre) should be applied prior to or at corn planting to meet the N needs of the crop for the first 30-45 days following emergence. Any additionally-required N can be applied as a sidedressing when the corn is 12 to 24 inches tall.

Soil testing in Virginia for available N before the growing season, as is practiced for other nutrients, does not accurately reflect the availability of N when it is most important to the crop (i.e., 30 to 45 days after emergence). The NO_3-N soil test procedure relies on sampling and testing after the crop has emerged and grown for several weeks. The concentrations of soil NO_3-N measured by this procedure are the result of many complex reactions affecting soil N and are more closely related to the need for supplemental N fertilization than any other procedure tested to-date.

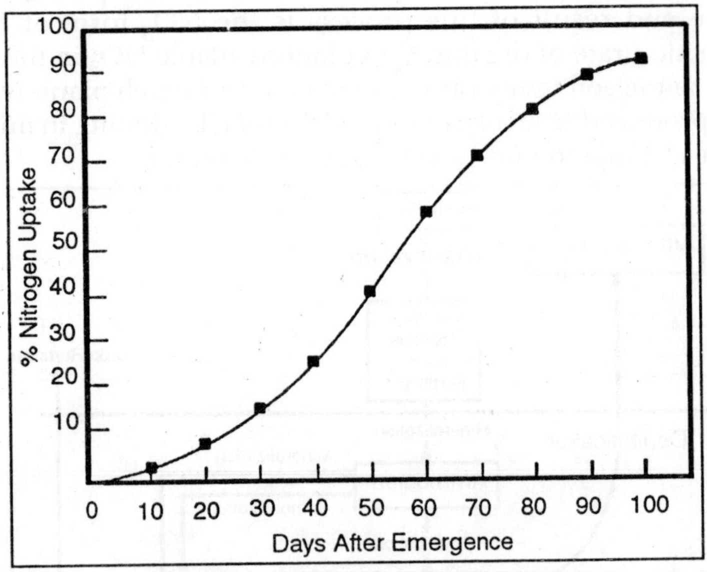

Fig. Nitrogen Uptake by Corn. (Origin)

PRE-SIDEDRESS SOIL NITRATE TEST

The pre-sidedress soil nitrate test (PSNT) is based on sampling of the surface one foot of soil after the soil has begun to warm and before the corn begins its most rapid growth rate (i.e., corn is 10 to 15 inches tall at the whorl). The amount of NO_3-N in the soil sample is an accurate index of plant-available N, and sidedress fertilizer N recommendations can be modified depending on the concentration of NO_3-N found in the soil.

Data from 47 field research experiments conducted in the Coastal Plain, Piedmont, and Ridge and Valley soil provinces in Virginia during 1990 and 1991 demonstrated that corn grain yields were maximized (i.e., relative yield = 1.0, or 100% of maximum) at soil NO_3-N concentrations above 18 parts per million (ppm). Enough N was mineralized, or made available from decomposing organic N, at the locations where NO_3-N concentrations were above 18 ppm to supply the seasonal N needs of the corn. Below 18 ppm, most of the relative yields were low, and supplemental N was needed to attain maximum yields. High relative yields occurred primarily where soil had received considerable contributions of organic N, such as from animal manures, biosolids, legume forages, or legume cover crops. Therefore, the PSNT should be utilized primarily on soils that have received significant amounts of organic N.

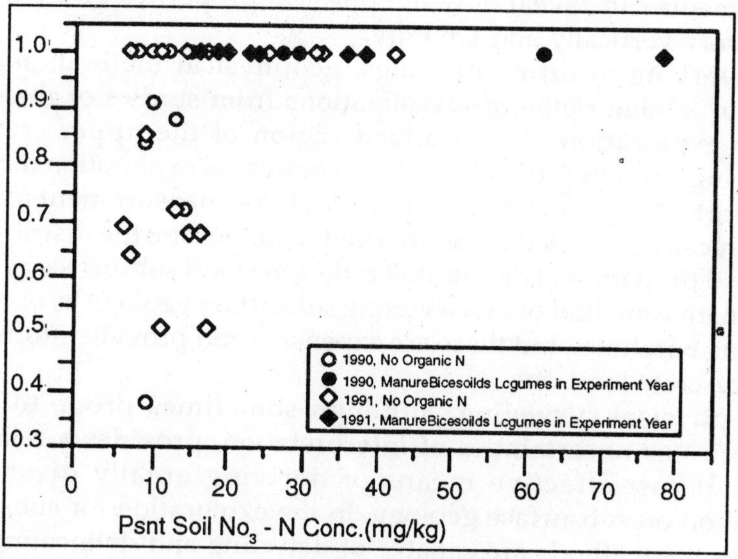

Fig. Relation Between Relative Grain Yield and Soil NO_3-N Concentration. (Origin)

Chapter 12

Limitations of Geophysics

This chapter is provided for readers with no prior knowledge of geophysical exploration methods and is pitched at an elementary level. It may be passed over by readers already familiar with the basic principles and limitations of geophysical surveying. The science of geophysics applies the principles of physics to the study of the Earth. Geophysical investigations of the interior of the Earth involve taking measurements at or near the Earth's surface that are influenced by the internal distribution of physical properties. Analysis of these measurements can reveal how the physical properties of the Earth's interior vary vertically and laterally.

By working at different scales, geophysical methods may be applied to a wide range of investigations from studies of the entire Earth to exploration of a localized region of the upper crust for engineering or other purposes. In the geophysical exploration methods (also referred to as geophysical surveying), measurements within geographically restricted areas are used to determine the distributions of physical properties at depths that reflect the local subsurface geology. An alternative method of investigating subsurface geology is, of course, by drilling boreholes, but these are expensive and provide information only at discrete locations.

Geophysical surveying, although sometimes prone to major ambiguities or uncertainties of interpretation, provides a relatively rapid and cost-effective means of deriving areally distributed information on subsurface geology. In the exploration for subsurface resources the methods are capable of detecting and delineating local features of potential interest that could not be discovered by any realistic drilling programme.

Geophysical surveying does not dispense with the need for drilling but, properly applied, it can optimize exploration programmes by maximizing the rate of ground coverage and minimizing the drilling requirement.The importance of geophysical exploration as a means of deriving subsurface geological information is so great that the basic principles and scope of the methods and their main fields of application should be appreciated by any practising Earth scientist.

THE SURVEY METHODS

There is a broad division of geophysical surveying methods into those that make use of natural fields of the Earth and those that require the input into the ground of artificially generated energy.The natural field methods utilize the gravitational, magnetic, electricai and electromagnetic fields of the Earth, searching for local perturbations in these naturally occurring fields that may be caused by concealed geological features of economic or other interest.

Artificial source methods involve the generation of local electrical or electromagnetic fields that may be used analogously to natural fields, or, in the most important single group of geophysical surveying methods, the generation of seismic waves whose propagation velocities and transmission paths through the subsurface are mapped to provide information on the distribution of geological boundaries at depth.

Generally, natural field methods can provide information on Earth properties to significantly greater depths and are logistically more simple to carry out than artificial source methods. The latter, however, are capable of producing a more detailed and better resolved picture of the subsurface geology. Several geophysical surveying methods can be used at sea or in the air. The higher capital and operating costs associated with marine or airborne work are offset by the increased speed of operation and the benefit of being able to survey areas where ground access is difficult or impossible.

A wide range of geophysical surveying methods exists, for each of which there is an 'operative' physical property to which the method is sensitive.The type of physical property to which a method responds clearly determines its range of applications. Thus, for example, the magnetic method is very suitable for locating buried magnetite ore bodies because of their high magnetic susceptibility. Similarly, seismic or electrical methods are suitable for the location of a buried water table because saturated rock may be distinguished from dry rock by its higher seismic velocity and higher electrical conductivity. Other considerations also determine the type of methods employed in a geophysical exploration programme.

For example, reconnaissance surveys are often carried out from the air because of the high speed of operation. In such cases the electrical or seismic methods are not applicable, since these require physical contact with the ground for the direct input of energy.

Geophysical methods are often used in combination. Thus, the initial search for metalliferous mineral deposits often utilizes airborne magnetic and electromagnetic surveying. Similarly, routine reconnaissance of continental shelf areas often includes simultaneous gravity, magnetic and seismic surveying. At the interpretation stage, ambiguity arising from the results of one survey method may often be removed by consideration of results from a second survey method. Geophysical exploration commonly takes place in a number of stages.

For example, in the offshore search for oil and gas, an initial gravity reconnaissance survey may reveal the presence of a large sedimentary basin that is subsequently explored using seismic methods. A first round of seismic exploration may highlight areas of particular interest where further detailed seismic work needs to be carried out. The main fields of application of geophysical surveying, together with an indication of the most appropriate surveying methods for each application. Exploration for hydrocarbons, for metalliferous minerals and environmental applications represents the main uses of geophysical surveying. In terms of the amount of money expended annually, seismic methods are the most important techniques because of their routine and widespread use in the exploration for hydrocarbons. Seismic methods are particularly well suited to the investigation of the layered sequences in sedimentary basins that are the primary targets for oil or gas.

On the other hand, seismic methods are quite unsuited to the exploration of igneous and metamorphic terrains for the near-surface, irregular ore bodies that represent the main source of metalliferous minerals. Exploration for ore bodies is mainly carried out using electromagnetic and magnetic surveying methods. In several geophysical survey methods it is the local variation in a measured parameter, relative to some normal background value, that is of primary interest. Such variation is attributable to a localized subsurface zone of distinctive physical property and possible geological importance.

A local variation of this type is known as a *geophysical anomaly*. For example, the Earth's gravitational field, after the application of certain corrections, would everywhere be constant if the subsurface were of uniform density.Any lateral density variation associated with a change of subsurface geology results in a local deviation in the gravitational field. This local deviation from the otherwise constant gravitational field is referred to as a gravity anomaly.

Although many of the geophysical methods require complex methodology and relatively advanced mathematical treatment in interpretation, much information may be derived from a simple assessment of the survey data. This is illustrated in the following paragraphs where a number of geophysical surveying methods are applied to the problem of detecting and delineating a specific geological feature, namely a salt dome. No terms or units are defined here, but the examples serve to illustrate the way in which geophysical surveys can be applied to the solution of a particular geological problem.

Salt domes are emplaced when a buried salt layer, because of its low density and ability to flow, rises through overlying denser strata in a series of approximately cylindrical bodies. The rising columns of salt pierce the overlying strata or arch them into a domed form.A salt dome has physical properties that are different from the surrounding sediments and which enable its detection by geophysical methods.These properties are:

- A relatively low density;
- A negative magnetic susceptibility;
- A relatively high propagation velocity for seismic waves;
- A high electrical resistivity (specific resistance).
- *The relatively low density of salt with respect to its:* surroundings renders the salt dome a zone of anomalously low mass. The Earth's gravitational field is perturbed by subsurface mass distributions and the salt dome therefore gives rise to a gravity anomaly that is negative with respect to surrounding areas. Figure presents a contour map of gravity anomalies measured over the Grand Saline Salt Dome in eastTexas, USA.

The gravitational readings have been corrected for effects which result from the Earth's rotation, irregular surface relief and regional geology so that the contours reflect only variations in the shallow density structure of the area resulting from the local geology.The location of the salt dome is known from both drilling and mining operations and its subcrop is indicated. It is readily apparent that there is a well-defined negative gravity anomaly centred over the salt dome and the circular gravity contours reflect the circular outline of the dome. Clearly, gravity surveys provide a powerful method for the location of features of this type.

- *A less familiar characteristic of salt is its negative magnetic:* Susceptibility, full details of which must be deferred to Chapter. This property of salt causes a local decrease in the strength of the Earth's magnetic field in the vicinity of a salt dome. Figure presents a contour map of the strength of the magnetic field over the Grand Saline Salt Dome covering the same area.

Readings have been corrected for the large-scale variations of the magnetic field with latitude, longitude and time so that, again, the contours reflect only those variations resulting from variations in the magnetic properties of the subsurface. As expected, the salt dome is associated with a negative magnetic anomaly, although the magnetic low is displaced slightly from the centre of the dome. This example illustrates that salt domes may be located by magnetic surveying but the technique is not widely used as the associated anomalies are usually very small and therefore difficult to detect.

- *Seismic rays normally propagate through salt at a higher:* Velocity than through the surrounding sediments. A consequence of this velocity difference is that any seismic energy incident on the boundary of a salt body is partitioned into a refracted phase that is transmitted through the salt and a reflected phase that travels back through the surrounding sediments. These two seismic phases provide alternative means of locating a concealed salt body.

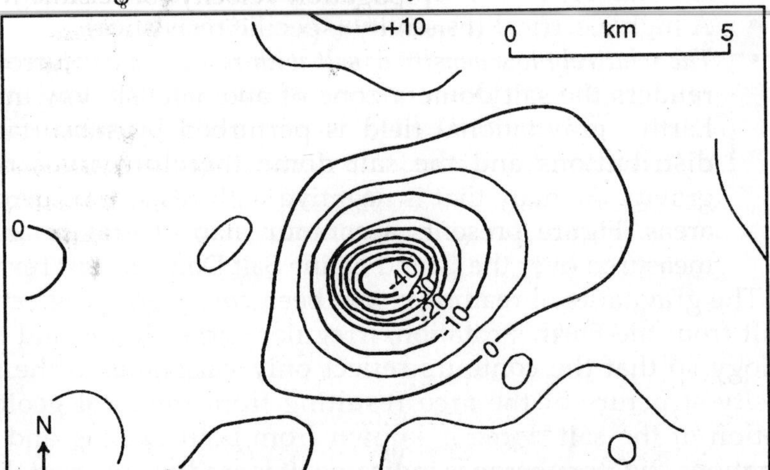

The Gravity Anomaly over the Grand Saline Salt Dome,.The Stippled area Represents the Subcrop of the Dome.

Fig. Magnetic Anomalies Over the Grand Saline Salt Dome,.The Stippled Area Represents the Subcrop of the Dome.

For a series of seismic rays travelling from a single shot point into a fan of seismic detectors, rays transmitted through any intervening salt dome will travel at a higher average velocity than in the surrounding medium and, hence, will arrive relatively early at the recording site. By means of this 'fan-shooting' it is possible to delineate sections of ground which are associated with anomalously short travel times and which may therefore be underlain by a salt body. An alternative, and more effective, approach to the seismic location of salt domes utilizes energy reflected off the salt, as shown schematically in Figure.

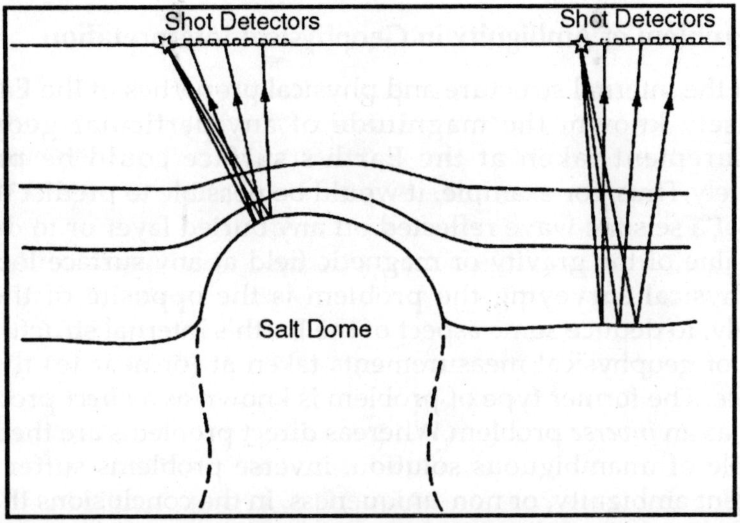

Fig. Simple Structural Interpretation of the Seismic Section, Illustrating some Possible ray Paths for Reflected Rays

- Earth materials with anomalous electrical resistivity may be located using either electrical or electromagnetic geophysical techniques. Shallow features are normally investigated using artificial field methods in which an electrical current is introduced into the ground and potential differences between points on the surface are measured to reveal anomalous material in the subsurface.

However, this method is restricted in its depth of penetration by the limited power that can be introduced into the ground. Much greater penetration can be achieved by making use of the natural Earth currents (telluric currents) generated by the motions of charged particles in the ionosphere. These currents extend to great depths within the Earth and, in the absence of any electrically anomalous material, flow parallel to the surface. A salt dome, however, possesses an anomalously high electrical resistivity and electric currents preferentially flow around and over the top of such a structure rather than through it. This pattern of flow causes distortion of the constant potential gradient at the surface that would be associated with a homogeneous subsurface and indicates the presence of the highresistivity salt.

Figure presents the results of a telluric current survey of the Haynesville Salt Dome, Texas, USA.The contour values represent quantities describing the extent to which the telluric currents are distorted by subsurface phenomena and their configuration reflects the shape of the subsurface salt dome with some accuracy.

The Problem of Ambiguity in Geophysical Interpretation

If the internal structure and physical properties of the Earth were precisely known, the magnitude of any particular geophysical measurement taken at the Earth's surface could be predicted uniquely.Thus, for example, it would be possible to predict the travel time of a seismic wave reflected off any buried layer or to determine the value of the gravity or magnetic field at any surface location. In geophysical surveying the problem is the opposite of the above, namely, to deduce some aspect of the Earth's internal structure on the basis of geophysical measurements taken at (or near to) the Earth's surface. The former type of problem is known as a *direct* problem, the latter as an *inverse* problem.Whereas direct problems are theoretically capable of unambiguous solution, inverse problems suffer from an inherent ambiguity, or non-uniqueness, in the conclusions that can be drawn.

To exemplify this point a simple analogy to geophysical surveying may be considered. In *echosounding*, high-frequency acoustic pulses are transmitted by a transducer mounted on the hull of a ship and echoes returned from the sea bed are detected by the same transducer. The travel time of the echo is measured and converted into a water depth,multiplying the travel time by the velocity with which sound waves travel through water; that is, 1500ms-1. Thus an echo time of 0.10 s indicates a path length of 0.10 ¥ 1500 = 150 m, or a water depth of 150/2 = 75 m, since the pulse travels down to the sea bed and back up to the ship. Using the same principle, a simple seismic survey may be used to determine the depth of a buried geological interface (e.g. the top of a limestone layer).

This would involve generating a seismic pulse at the Earth's surface and measuring the travel time of a pulse reflected back to the surface from the top of the limestone. However, the conversion of this travel time into a depth requires knowledge of the velocity with which the pulse travelled along the reflection path and, unlike the velocity of sound in water, this information is generally not known. If a velocity is assumed, a depth estimate can be derived but it represents only one of many possible solutions. And since rocks differ significantly in the velocity with which they propagate seismic waves, it is by no means a straightforward matter to translate the travel time of a seismic pulse into an accurate depth to the geological interface from which it was reflected. The solution to this particular problem, is to measure the travel times of reflected pulses at several offset distances from a seismic source because the variation of travel time as a function of range provides information on the velocity distribution with depth.

However, although the degree of uncertainty in geophysical interpretation can often be reduced to an acceptable level by the general expedient of taking additional (and in some cases different kinds of) field measurements, the problem of inherent ambiguity cannot be circumvented. The general problem is that significant differences from an actual subsurface geological situation may give rise to insignificant, or immeasurably small, differences in the quantities actually measured during a geophysical survey. Thus, ambiguity arises because many different geological configurations could reproduce the observed measurements. This basic limitation results from the unavoidable fact that geophysical surveying attempts to solve a difficult inverse problem.

It should also be noted that experimentally-derived quantities are never exactly determined and experimental error adds a further degree of indeterminacy to that caused by the incompleteness of the field data and the ambiguity associated with the inverse problem. Since a unique solution cannot, in general, be recovered from a set of field measurements, geophysical interpretation is concerned either to determine properties of the subsurface that all possible solutions share, or to introduce assumptions to restrict the number of admissible solutions. In spite of these inherent problems, however, geophysical surveying is an invaluable tool for the investigation of subsurface geology and occupies a key role in exploration programmes for geological resources.

Index

G

H

I

K

L

M

T

U

V

W